Ironworks on the

Saugus

IRONWORKS ON THE

Saugus

*The Lynn and Braintree Ventures
of the Company of Undertakers of the Ironworks
in New England*

BY E. N. HARTLEY

NORMAN: UNIVERSITY OF OKLAHOMA PRESS

International Standard Book Number: 0–8061–0366–3 (cloth);
International Standard Book Number: 0–8061–0957–2 (paper)
Library of Congress Catalog Card Number: 57–5956

Copyright 1957 by the University of Oklahoma Press,
Publishing Division of the University.
Manufactured in the U.S.A.
First edition, 1957.
Second printing, 1971.

To Dexter

Per Omnia

Preface

THIS BOOK IS a history of the Lynn and Braintree ventures of the Company of Undertakers of the Ironworks in New England and of the making of iron in America through the seventeenth century. It is an outgrowth of work on the Saugus Ironworks Restoration, a rather unique venture in industrial, civic, and scholarly co-operation in which it was my privilege to serve as research historian from the spring of 1949 to the autumn of 1954.

Here a large modern industry, organized in American Iron and Steel Institute, sought to commemorate its founders by setting up a faithful replica of "Hammersmith," the mid-seventeenth-century plant in which the successful, sustained, and integrated production of cast and wrought iron was first achieved within the limits of the United States. The project was carried in the name and legal title of the First Iron Works Association, Inc., of Saugus, Mass., a predominantly local group of interested citizens. In the work of research and reconstruction many people here and abroad, representing various fields of business endeavor and a number of scholarly disciplines, fused their skills and interests. All contributed to my own efforts, and therefore to this book. I regret that properly to acknowledge their help, and to exonerate them from responsibility for the use I have made of that which they freely offered, is rendered nearly impossible by the size and degree of generosity of the company.

vii

I cannot fail publicly to express my thanks to four groups with which I was in close contact over many months. American Iron and Steel Institute provided me with splendid resources and warm encouragement. To its officers and staff, in particular to Max D. Howell, executive vice president, and George S. Rose, secretary, and to its Committee on General Research, I am deeply indebted. The First Iron Works Association, especially in the person of its president, J. Sanger Attwill, has been most kind and has demonstrated a commitment to accuracy in the reconstruction and in the historical research which has never wavered. It gave me access to its collection of historical materials, and its clerk, Miss M. Louise Hawkes, shared with me her rich knowledge of local history and genealogy, and spent many hours transcribing timeworn manuscripts. Her gentle spirit has been an inspiration to me and to all who worked at Saugus. To the Restoration architects, the Boston firm of Perry, Shaw and Hepburn, Kehoe and Dean, and in particular to its able representative Conover Fitch, Jr., I am indebted for data, interpretations of data, and many happy hours of joint effort in bridging the gap between the raw materials of technological history and the various units of an actual ironworks. Architects and historian profited alike from the practical skills of the builders of the Restoration, H. M. Bogart and Company, of Charlestown. Albert W. Barnard, a partner in the firm, and Donald P. Jones, job superintendent, taught me much.

Complementing the historical research in the gathering of data with which to authenticate the Restoration was the archeological work of Roland Wells Robbins and his assistants. From his findings I derived many new insights and not a few corrections of old ones. In checking his and my data, the resources of several steel company laboratories were invaluable. I am in the debt of a number of their technical specialists, especially of Dr. Hobart M. Kraner, of Bethlehem, and Earle C. Smith, of Cleveland. In the preparation of publicity materials and preliminary historical summaries, Hill and Knowlton, Inc., the public relations counsel for the Steel Institute and for the Saugus Restoration, in particular as represented by Howard Stephenson, a former vice president, has been most helpful.

Professor Samuel Eliot Morison, who first sparked my interest in colonial American history, has given me encouragement and help in the present study. I am especially grateful for his having made available his notes on business and industry in seventeenth-century Massachusetts. In the field of ironworks history, to which I came as a novice, I have derived much from the counsel of Professor Arthur C. Bining, of the University of Pennsylvania, dean among scholars of American iron, and of the late Charles Rufus Harte, of New Haven, historian of the early Connecticut iron industry and a man richly versed in American industrial history in general. Dr. H. R. Schubert, historical investigator to The Iron and Steel Institute (British), served as consultant on the Saugus Restoration and has been quite helpful to me. In the course of his Saugus visit he gave us much good advice based on years of research in the history of iron in the British Isles and on the Continent. He has also made available a profusion of documentary materials, scholarly articles, and friendly criticism. In my research and writing, and in other ways, I have also benefited from the friendship, special competences, and interests of present and former colleagues at M.I.T., including Professors John M. Blum, Alfred D. Chandler, Karl W. Deutsch, William C. Greene, Arthur Mann, Elting E. Morison, John B. Rae, Walt W. Rostow, Lawrence W. Towner, and C. Conrad Wright.

The research in the English documentary sources was carried through in exceptionally able fashion by Brian W. Clapp, a doctoral candidate at the London School of Economics at the time he was recommended to me by Professor T. S. Ashton. The latter was most helpful to him, as were Mr. F. G. Fisher, of the same institution, Mr. T. F. Reddaway, of University College, London; Mr. Hugh S. Pocock, clerk to the Merchant Company; the library staffs of the London School of Economics, the British Museum, and the Bodleian; and the staff of the Public Records Office. Mr. Clapp must bear no responsibility for inaccuracies of citation or unsound interpretations which may have crept into my text as I worked from his carefully prepared and beautifully written research reports. The measure of my indebtedness to him will be manifest to any reader.

Here in Cambridge I was helped, successively, by two able and energetic research assistants, Mrs. Marcia Hollender Kendall and Mrs. Mary Lou MacElhose Joyner. Both handled with grace and dispatch research which was often as taxing as dull. Both digested and summarized materials of no inconsiderable complexity. Mrs. Joyner, whose term of service was longer, saved me from many a lapse of judgment and statement, and also typed most of the manuscript. In trying to square gleanings from old documents with the topography and geology of the Saugus, Braintree, and Quincy areas, I was assisted by Drs. Laurence LaForge and J. B. Thompson, Jr., geologists at Harvard. Attorneys Laurence F. Davis and G. Marsden Leonard, of Lynn and Framingham, respectively, ran careful surveys of land titles in the Saugus and Braintree-Quincy areas which shed much light on the location and extent of the landholdings of the old ironworks. Genealogists who compiled information on families with ironworks connections have also been of assistance.

My work, and that of the young ladies who helped me, has been made more pleasant and more efficient by the courteous and skillful staff members of a great many libraries and manuscript collections. Among those to which we are especially indebted, whether for permission to photostat or for aid in collecting data, or both, are: the M. I. T. Library; Widener, Houghton, and Baker libraries, at Harvard; the Boston Public Library; the Boston Athenaeum; the New England Historical Genealogical Society; the Massachusetts Historical Society; the American Antiquarian Society; the Old Colony Historical Society; the Connecticut State Library; the New York Public Library; the New York Society Library; the Library of American Iron and Steel Institute; the Monmouth County (N. J.) Historical Association; the Library of Congress; the Virginia State Library; the Royal Society, London; and the Jernkontoret Library, Stockholm. Our impositions on those responsible for the Massachusetts State Archives, the Archives of the Supreme Judicial Court of Massachusetts, and the court records of Essex, Suffolk, Middlesex, and Norfolk counties, have been conspicuously heavy. The kindness with which they have been received has been extraordinary. I am especially grateful to the Harvard University Library for permission to reproduce illustrations from *Descriptions des Arts et Métiers*, the collection given to Harvard College in 1789 by John Adams, Vice President of the United States.

During much of the time in which I was working on the restoration project and on this book, I had been relieved of some of my teaching duties at M. I. T. For arranging this, my thanks go to Professors Howard R. Bartlett, chairman of the Department of Humanities, and John E. Burchard, dean of the School of Humanities and Social Studies.

Finally, I owe more than I can here say to Walter S. Tower, former president, and Charles M. Parker, assistant vice president, of American Iron and Steel Institute. Together with the late Quincy Bent, chairman of the Reconstruction Committee of the Steel Institute and the First Iron Works Association, and a valued friend, they launched me in the venture and have stalwartly championed my efforts down to the present day. To them, but for the higher claims of my good wife, this book would have been dedicated. For much of such merit as it may possess they are responsible.

E.N.H.

Hayden Memorial Library
Massachusetts Institute of Technology

Contents

Illustrations

IRONWORKS ON THE SAUGUS

Ironworks on the

Saugus

I

Prototype of American Industry

WITHIN TWENTY YEARS of the settlement of Boston in 1630, enterprising men had erected large integrated ironworks in the midst of a virgin wilderness only gradually succumbing to agriculture. Promoted by John Winthrop the Younger, they were the property of the Company of Undertakers of the Iron Works in New England, a group of some twenty Englishmen and three or four Massachusetts residents. With their money and the blessing and concrete support of the local authorities, Winthrop and his successor, Richard Leader, built between 1644 and 1647 two plants, one a few miles north and the other south of Boston, to convert bog iron into cast and wrought iron.

There had been earlier industrial activity in Massachusetts, activity which turned Nature's gifts into products which both the young Colony and the Mother Country could use. These ironworks, however, were different in that they were large-scale factory enterprises involving joint financing, a complicated technology, specially imported workmen, and heavy capital risk. They were big business and heavy industry. In the Bible Commonwealth they stood out as atypical, anachronistic and wonderful.

And so they were regarded by the Puritans of mid-seventeenth-century Massachusetts. To the noisy, dirty plants on Saugus River in Lynn and on Furnace Brook and Monatiquot River at Braintree flocked sightseers drawn as to a new world's wonder, and to the em-

3

barrassment of the Company agent who had to entertain at least the more important of them. The Undertakers, assuming that popular curiosity should have been satisfied by the spring of 1652, forbade John Gifford, their third agent, to continue his predecessor's practice.[1] Progress toward completion of the ironworks had been eagerly watched by the Governor whose son had promoted them. He passed along reports on production to his namesake, whose connection with the ironworks he inspired was short-lived. When Governor Winthrop died and a committee of the General Court examined his papers, it listed, uniquely in the midst of documents concerning affairs of state, a category of "Writings conceȓ the iron workes," testifying to their importance to Winthrop and the community over which he presided.[2] The chroniclers of settlement, Winthrop, Hubbard, and Johnson, all made reference to the undertaking which had been set afoot with great expectations.[3]

Any student of colonial history knows that there were dozens of loudly hailed grand designs and projects, most of which came to naught and stand embalmed in the records only as evidence, now quaint and touching, now grandiloquent beyond belief, of the dreams of pioneers. The New England ironworks, however, became realities, and impressive realities, despite the limited duration of their effective production. One of the Company's agents adjudged them "as good as any worke England doth afoarde."[4] The researches of modern scholars do not seriously weaken his claim. At Braintree, scene of faltering beginnings, was a blast furnace of good size, a forge consisting of finery, chafery, and giant water-power-driven hammer, a charcoal house, and workmen's cottages. At Lynn, there was a complete ironworks, whose design and engineering were as bold as sophisticated. Here was a huge furnace, a forge comprising two fineries, a chafery, and a big hammer, an extensive water-power system, good storage facilities, workmen's accommodations, and a pier for the use of the small boats which plied the Saugus River laden with the ironworks products. Here was a rolling and slitting mill, the first in the

---

[1] Ironworks Papers, 30.

[2] *Mass. Records,* III, 179.

[3] John Winthrop, *History of New England* (James Savage, ed., 2 vols., Boston, 1843), II, 261; *Johnson's Wonder-Working Providence* (London, 1654) (J. Franklin Jameson, ed., New York, 1910), 73, 245–46; William Hubbard, "General History of New England," 2 *M.H.S. Coll.,* VI, 374.

[4] S. P. Col., XV, fol. 43 (Gay Transcripts [M.H.S.] I, 33).

New World, and set up when there were only about a dozen of which we have record in the British Isles and on the Continent.[5] To build all this had taken the willingness to risk of capitalists, the vision of men we today call engineers, the sweat of all but unknown workmen achieving performances of high skill in the working up of timber and stone and iron.

Hammersmith, as the plant at Lynn was called, was, in the dreams of the men who conceived it, and to a fair degree in fact, a self-sufficient economic unit and an articulated ironworks in which raw materials were carried through to finished product with a flow which a modern industrialist could but admire. Braintree works was smaller; indeed, during most of its operating period it served as an ancillary forge joined with the larger one at Lynn in converting the cast iron from Hammersmith Furnace into merchantable bar iron. It was, nevertheless, far from a crude establishment. Joined to the producing units, in time, was a warehouse and factor's office at Boston which handled the procurement of such things as the ironworks could not produce themselves and the sale and export of much of their output. The whole seems strangely modern in conception and execution.

Activities at Braintree and Lynn first required the searching out, by means of a "survey," of suitable ore deposits and likely plant locations. Part of this was a matter of "science" following in the path of rumor. It had been Winthrop's tracking down of reports of mineral finds that provided the impetus for a trip to England to raise capital. When he returned successful, a similar job was undertaken more systematically. Part of it was a matter of "engineering" in which availability of water power, of means of transportation, of workers' housing, and of supplies, and accessibility to markets were all carefully considered. Obviously, the work was carried out with far simpler and less reliable techniques than are used in the location surveying of a modern American corporation. In essence, however, Winthrop was doing the same job and asking many of the same questions as would, let us say, the Du Pont Company in deciding on the location of a dacron plant.

Capital raising was perhaps more, rather than less, complicated than it is today. The appropriate techniques were less well systematized. Basically, however, it was a matter of selling a proposition

[5] Dr. H. R. Schubert, Historical Investigator, Iron and Steel Institute (British), in address to First Iron Works Association, Inc., Saugus, Mass., June 28, 1952.

to a group of men who had money to invest. Once persuaded by the promoter, Winthrop, these men had to organize as a company, in a form deemed appropriate and feasible. Since the company was to be separated from its investment, the latter had perforce to be expended under the direction of paid managers, and accounting and other control techniques worked out. There were precedents of a sort in the ironworks which had been set up in Ireland and elsewhere by a burgeoning Mother Country industry running short of fuel and tempted, perhaps, by cheap unskilled labor. The Atlantic Ocean, however, is broader than the Irish Sea. Many of the troubles of the Massachusetts ironworks ultimately went back to less than perfect solution of these accounting and control problems. In the surviving court papers which document them are good data for the study of a groping toward forms and procedures which are now commonplace.

Since the ironworks were erected in a colony, and with the intervention of its magistrates, their story is rich in materials on the relations between government and industry, and at the level of both local and English authority. The home government played no part in the establishment of the ironworks. Nevertheless, their destiny became entwined in colonial politics, in the relations between the Puritans who had emigrated and those who had stayed at home eventually to rise to power with Cromwell. Within Massachusetts there were first the working out of arrangements which would tempt capitalists to invest in this infant industry and still safeguard the interests of the settlers, and then the policing of these arrangements. Eventually, there was to arise the necessity for finding and carrying out justice as the whole ironmaking establishment ran into legal snarls which would have given Solomon himself pause. Here are materials on relations between "state" and "business," in specific and occasionally complicated situations. Interesting as data, they might tempt some to generalize and form value judgments—doubtless according to their politics—on the policies worked out and followed.

An industrial establishment is also, of course, the product of the labors of men. The employees of the Company of Undertakers formed a special group. Thanks to the deep-rooted customs and procedures peculiar to their trade, they would have been different from their neighbors in the Mother Country. In mid-seventeenth-century Massachusetts they were all but unique. Interactions between the special group and the general community provide some of the most significant facets of the Hammersmith story. Within the plant were many of the personnel problems with which men wrestle today, the recruiting of relatively scarce specialists, the providing of on-the-job training for younger men, the driving of wage bargains, the discipline of recalcitrants. And then, to look at it from the other side of the coin, these were individuals, now marrying and having children, now working and playing together, now rising in the social and economic scale, now down on their luck. We know far more of their lapses, moral and financial, than we do of the more positive elements in their activities. There is, however, in the Hammersmith story a cameo of the life of early American industrial workers which is good and useful despite its uneven relief.

Finally, these were ironworks. The making of iron is an extended and reasonably burdensome operation. To make it in a virgin settlement required the setting up of large units and heavy machinery. It called for both the importation of techniques and, in some measure at least, their adaptation to the local physical environment. The technology, however, was, again, but a part of a larger whole, a business. The story of the ironworks at Lynn and Braintree is in large part a business history with data on organizational forms, managerial control devices, production costs, sales, profit, and losses, and, of course, the bookkeeping on these latter items. There are also materials, less good than one would wish, bearing on marketing and product distribution.

Broadening our focus still further, these ironworks were part of a whole community. If there was interaction between business and government, so there was interaction between the ironworks people and their neighbors. Some of the latter joined in the less specialized work on a part-time basis or in off seasons. There was considerable intermarriage. All went to the same churches—and to the same taverns. It has been said that the poor have no history. Nevertheless, it is impossible to believe that these little people, mingling with their farmer neighbors in the hundred contexts of a frontier settlement, failed to influence the shape of the whole community, effecting change as they were themselves changed. Indeed, it is reasonably clear that in encouraging the setting up of ironworks the Puritan magistrates were opening a Pandora's box, whose contents would undercut many of the things for which they stood.

In terms alike of its success or failure and of its historical significance, the relationship between a given enterprise and the society and economy it serves is crucial. Quantitative data on the economy of Massachusetts and New England in the seventeenth century are not good. Questions of scale of enterprise, of ratio of productivity to demand and of wages to cost of living are largely unanswerable. Thanks in part to the fact that the Undertakers' plants were built on a rather grander scale than the general economy could support, at least as long as the proprietors demanded hard money for their products, they got into serious financial difficulties less than ten years after they were first conceived. These difficulties, of course, were richly aired in the courts. In the judicial records, which Massachusetts has well preserved, are to be found at least the crude raw materials of quantitative economic data on an important industrial venture in its total economic milieu. Though we cannot handle them with the rigor they deserve, we are not reduced to a mere study of the anatomy of a pair of ironworks or to the mere chronicling of their rise and fall.

Within the ironworks and in their total surroundings nothing was static. All of the institutional and environmental problems had backgrounds, all led to what men hoped were better techniques of solution. At any particular stage, however, they were being handled by men in whose attitudes were some of the past, much of the present, and a sprinkling of the future. Industry cannot stand still. This seventeenth-century industry did not. Continuity greatly outshines the discontinuities in its story, whether we look backward to the roots from which it sprang or forward toward the development that has taken place between its day and ours. In what went on at the ironworks at Lynn and Braintree, there was far more of the "old" than of the "new." There were also, however, some American "firsts."

Though zealous local historians early claimed that each of these ironworks was the "first" in America, neither merits the designation if by the term is meant no more than a place where iron was made. Braintree Furnace antedated Hammersmith. Both were antedated by small-scale but recorded and successful attempts to reduce American iron ores to useful metal in Greenland, in Canada, and in Virginia. Braintree Furnace was a failure though it did make iron. Braintree Forge became a branch plant of Hammersmith. The latter plant, despite its predecessors, can fairly be called America's first successful integrated ironworks. Its centrality in this book hinges,

8

however, less on its primacy than on its scale, its catalytic role in Puritan New England, its function as a training school for men who went on to other places and built up a fair share of the colonial American iron industry. In all this it overshadowed Braintree Works and all the other earlier but less significant ironmaking ventures.

These plants of the Company of Undertakers can also claim a kind of "first" for the northern colonies in the seventeenth century as the earliest large-scale industrial enterprise with a heavy fixed investment. The Company missed by a narrow margin, apparently, a further claim to fame, this time as the first industrial corporation in English America. Viewed through narrow legal eyes, the Undertakers made up a partnership. The formal requirements of incorporation were lacking, probably because not only were the times not yet ripe for the chartering of private industrial companies, but for the Massachusetts Bay Company to have done so would have been an almost flagrant violation of jealously guarded English sovereign rights. Nevertheless, in structure and actions, the Company partook of many of the characteristics of a corporation. Certainly, it was a fascinating transition between the partnership and the legal form without which heavy, long-term investment—and modern capitalism—would be impossible.

There was at least some novelty, too, in the workers' situation. It was the financial ambitions of certain men, not their religious faith, which had taken them to Massachusetts. They were, in a real sense, an alien element. As such, they are prototypes of foreign population nuclei set down in the midst of settled residents—nuclei which have played so important a role in American industrial development in the colonial period and right down to today. Despite the fact of a common language, there was as much of a problem of "assimilation" for these ironworkers and their families as for our French-Canadian millworkers or the Slavs of the steel industry of some decades ago.

Hammersmith was in many respects a first American "company town." We have said that the capitalists aimed at self-sufficiency. They wanted a unit which would not only make iron but grow its own food, and otherwise keep its dependence on the outside to a minimum. It was an embryonic version of what Professor Bining, able scholar of the history of American iron and steel, has well named the "iron plantation."[6] There was in all likelihood a company store. Cer-

[6] Arthur Cecil Bining, *Pennsylvania Iron Manufacture in the Eighteenth Century* (Harrisburg, 1938), 29 *et seq.*

tainly there was company housing. But for all this, the ironworks people did not live in isolation. They were thrown into contact with their Puritan neighbors. At first, and for some time, there was a mighty cleavage between the hard-drinking and stubbornly independent workers and those who had come to Massachusetts primarily for religion's sake. Eventually, however, the workers fused with the total population, many of them becoming church members and freemen, all but indistinguishable from the Puritans, for whom their tough refractoriness had once been a heavy cross. A similar process has been repeated a hundred times and in a hundred places in subsequent American history.

In the field of technology the Company of Undertakers can claim important first transplantings and at least one major innovation, or improvisation. The "priority" of the slitting mill has already been mentioned. The date of construction of the forge at Braintree and Lynn is so obscure that it is impossible to tell which came first, but one or the other was doubtless the first example in America of the Walloon process forge, which used both finery and chafery hearths for iron refining. Braintree Furnace achieved the first recorded and successful production of cast iron within the limits of what is now the United States. It or Hammersmith Furnace can claim a significant innovation in the matter of fluxing agents. This was the use of a gabbro found on the peninsula of Nahant as an effective substitute for the limestone in common use in England, and for the shells which were used as flux in many American furnaces in the colonial period. Here there were no precedents. The potentialities of the gabbro can only have been discovered by trial and error tactics, probably initiated in the belief that the dense igneous rock was an iron ore.

Various individuals among the ironworks people have been, or deserve to be, awarded "firsts" of their own. John Winthrop the Younger has often been called America's first scientist, eminent enough to be admitted to membership in the Royal Society in London, to which he contributed papers. His successor, Richard Leader, has considerable claim to the title of America's first engineer. It was he who built Hammersmith, and Hammersmith called for engineering of high order. He also erected large sawmills in the present town of Berwick, Maine, which were the wonder of their time and place, grand enough to give the Great Works River the name it has carried down to our own times. An employee named Joseph Jenks, a skilled blacksmith soon turned independent or semi-independent entrepreneur, got the first machine patent awarded in America for "engines of mills to go by water," and another patent for an improved kind of scythe. He has also been credited with having made the dies for the pine tree shilling, and named as the builder of the first fire engine in America, although on highly doubtful authority in both instances. Excavations in the area of his little shop at Hammersmith suggest that he may well have had the first hammer- or plating-mill for fabricating iron in the colonies, and he was probably the first to make brass pins and brass platings as well.

The lines of continuity are far more important, however, than all these "firsts." First of all there was, of course, technological continuity. The works which arose at Lynn and Braintree were heirs of a long tradition. It was customary among nineteenth-century writers on the history of iron to begin their story with the Iron Age and carry it down faithfully through Biblical times, classical antiquity, and the medieval world, or to trace out the various stages of evolution from the crude bloomery, still to be seen in primitive cultures, to the blast furnace, and through to the various ironmaking techniques prevalent in their own day. We need not do likewise here, but the choice of the indirect process by the men who set up the Braintree and Hammersmith works calls for at least a brief summary of the immediate technological background.

The first working American blast furnaces were copies of those in use in the England from which their builders came. Originating in the Low Countries, in the region of Namur and Liége, and thence transplanted to France, they had been introduced into England in the last decade of the fifteenth century. There they multiplied until England had a flourishing iron industry, so flourishing as to raise alarm, as early as the middle of the sixteenth century, about its consumption of timber for charcoal. One main impetus to colonial ironworks was the hope that the virgin forests of the New World would supply the vast quantities of fuel used in the reduction of ore to iron.

The indirect process of iron manufacture is more complicated than the direct. The latter calls only for the use of the bloomery, a hearth in which, in a single if discontinuous operation, wrought iron is made directly from the ore. In the indirect process, the blast furnace smelts the ore and turns out cast iron in the form of sows and pigs. By the Walloon method, these are then converted into wrought iron in a series of heating and hammering operations in the forge with its two types of hearth, the finery and the chafery, and its big water-power-driven hammer. While blessed with many advantages in actual operation, the indirect process requires a much heavier investment than the technique by which men had made iron from time immemorial. America's iron ore resources could have been handled by the bloomery process. Indeed, many later colonial ironmaking ventures used the bloomery. It is significant that when the Company of Undertakers launched its Massachusetts operations it chose to adopt the indirect process at both Braintree and Lynn, to use the best technique for making iron then available in England and elsewhere. Dissenting opinion on the soundness of the choice is taken up in a later chapter.

Working practice at the ironworks was traditional, except to the extent that peculiar local conditions may have made minor adjustments necessary. It was traditional in the literal sense of having been passed along from workman to workman. Men did what their fathers had done, wrapping their activities in a cloak of trade secrecy and fending off the inquiries of outsiders, regardless of who they might be—curious scientists or scholars or even their employers. There is a corpus of technical writings of the late medieval and early modern period. But its contents are thin indeed, so far as the processing of iron goes. While, therefore, New England ironworks got started with a well-developed technology, it was largely in the heads and hands of the workmen. These were imported specialists, functioning in a manner analogous to that of Frenchmen and others who had carried the indirect process to England. What they had got from their fathers they passed along to their sons, with consequences which we shall shortly see.

There was also a deep background for the commercial operations in use at plants and warehouse, the process of capital accumulation, the form of capitalistic organization, and the various fiscal and control techniques. To survey this background would carry us into such fields as the development of double-entry bookkeeping, of merchant partnerships and joint-stock companies, of attorneyship, agency, and factorage. Again, there is no need here to outline fields in which so much excellent, scholarly work has been accomplished. In the surviving data on the activities of the Company of Undertakers, there is ample evidence of the following of standard practice along with the searching out of new techniques for running a costly industrial operation at great distance and under conditions of slow and uncertain communication.

In the background, and not so far in the background, of the two plants stood the "state" in the persons of the magistrates of Massachusetts. For the intervention of the state in economic activities in the middle of the seventeenth century there was ample precedent. On one side was the whole range of policies which we lump under the heading of mercantilism, that system by which a nation sought, by stimulus and regulation, to control economic activity toward its own greater well-being. Massachusetts was a colony, not a nation; nevertheless, the men who led it carried out mercantilist policies in many areas. On the other side were policies deriving largely from medieval notions as to the relations between government and economics. In a blending of these the General Court had seen fit to fix prices and wages, impose quality controls, require Massachusetts residents to work, and pass sumptuary legislation. Fortunately, in many cases, measures adopted to achieve a better measure of economic self-sufficiency squared nicely with the plans for a Christian community in which the magistrates, God's anointed, led sinful men along the lines of such precepts as "The whole is greater than the sum of its parts," "In what you have you are God's steward," "Life is but a fleeting moment in eternity." Occasionally, there was conflict between the two wellsprings, and the conflict was heightened by divergences between the old, which was Catholic and medieval, and the new, which was Protestant and modern.

The cleavage is nicely illustrated in the magistrates' handling of the new ironworks. To develop a local natural resource and make directly available a basic metal vital in the development of a well-populated settlement, the General Court granted the Company of Undertakers land, exemption from taxes and military service, and even a monopoly on the making of iron within its jurisdiction. At the same time, however, it chose to impose restrictions, conspicuously a ceiling price on bar iron, in the interests of the community at large. The Colony's leaders were aware of the often unfortunate consequences of English grants of monopoly and other subsidy to infant industries. They were also aware that capitalists, who bore, like all men, the taint of the sin inherited from Adam's fall, needed to be regulated lest the very profit motive which would give Massachusetts its iron do injury to its settled residents. In the long-drawn-out negotiations for the various grants and immunities is the record of basic attitudes in flux in the transition from a medieval to a modern world. These are the more important roots of the Hammersmith and Braintree ironworks. The roots are deep. As we study them, we see signs and portents of the future. Now and again, as in the area just mentioned, there is a mingle-mangle of old and new. If we reverse the direction of our scrutiny and look at the ironworks against that which followed them, we find, once more, continuity and discontinuity. Samuel Eliot Morison exaggerated at least a little in the judgment he hazarded in his fine *Builders of the Bay Colony:* "If the history of all the local ironworks and tool factories established in New England before 1800 were followed up, it would probably be found that most of them can be traced through Jencks and the Leonards to the old Hammersmith plant at Saugus."[7] Some, but less than the majority, of the later ironworking establishments in colonial New England got built by or employed men who had once worked at Hammersmith and Braintree, or their descendants. Between Hammersmith and the later plants there was also, at least for many decades, a marked difference in scale.

Most of the ironworks which followed the plants of the Company of Undertakers were small, locally financed, and, in the great majority of cases, direct-process or bloomery plants. This holds for both those whose continuity with Hammersmith and Braintree can be documented and those set up and worked by men having no known association with these pioneer establishments. Hammersmith, in particular, was bigger and better than most of its children, or the rivals of its children. This is in no way strange. Hammersmith was in many respects an artificial creation, conceived in the grand manner, lushly financed, and richly supported by governmental authority. The birth of the children and the others was more natural, the products healthier though smaller, or perhaps healthier because smaller, in better scale with the economy they served.

In those plants which did go back to Braintree and Hammersmith, we are dealing with a continuity of personnel, not of capital proprietorship. While the Company of Undertakers operated successfully it needed all the skilled help it could get. Since it had a

7 Samuel Eliot Morison, *Builders of the Bay Colony* (Boston and New York, 1930), 279–80.

monopoly, there was by definition no place within the limits of Massachusetts to which its people could move on and ply their trades. When it foundered on the shoals of indebtedness in 1653, its creditors tried to carry on but, though they expended no small amount of capital, operations were fitful and, on the whole, unsuccessful. At various periods of slump, conspicuously in that following the catastrophes of 1653, workmen were unemployed, in a sorry plight, and eager to seek out employment elsewhere. The establishment of ironworks in Plymouth Colony and in Connecticut, and at new sites in Massachusetts following the cancellation of the Undertakers' monopoly, gave them their opportunity. By 1700 no less than eight ironmaking plants had been set up in Massachusetts, Connecticut, Rhode Island, and New Jersey. All but one of these were started or staffed, at least in part, by men who had once worked at Braintree and Lynn. The spreading out of the sons and grandsons of one-time employees of the Company to smaller iron-fabricating shops, hammer mills, and smithies was more impressive still.

It was largely thanks to John Winthrop the Younger that an ironworks was set up at New Haven in 1655–56. This plant was largely staffed by the "unemployed" of Hammersmith. John Gifford, the Undertakers' third agent, went on eventually to set up a second ironworks at Lynn. The son of Joseph Jenks was for a while connected with a bloomery at Concord, in whose bill of privileges from the General Court appeared the liquidation of the Undertakers' monopoly. He later set up a small plant at Pawtucket, Rhode Island. Other Jenks sons and grandsons stayed in the family trade of blacksmithing and practiced their skills in a number of places. The Leonards, the second family mentioned by Professor Morison, fanned out to Taunton, Rowley Village, and Tinton Falls, New Jersey, even in the first generation. Doctor Fobes, the author of a *Topographical Description of the Town of Raynham* (Massachusetts), wrote in the eighteenth century: "Where you can find iron works there you will find a Leonard."[8] The quotation, cited by almost every historian of the American iron industry, has considerable validity, and for more than the area of Bristol County, to which the old Doctor had particular reference. At these and other plants also appear the names of less well-known and less successful people than Jenkses and Leonards, people with whom they had once worked side by side at Hammersmith and Braintree Forge.

Besides all of these, however, at the plants just cited and at others set up in the same general period, there were workers and managers who seem never to have been associated with the works of the Company of Undertakers. Where they got their training is quite obscure. Perhaps they were newcomers, immigrants who came to these shores as anxious to find a profitable sphere for the exercise of their trades as the men who sooner or later, willingly or perforce, left Lynn and Braintree. Precise statistical data are lacking. By rough estimate, however, the sons of Hammersmith, and their sons and grandsons, were outnumbered by ironworkers who brought their skills from other places. This is indisputable for the colonies in general. It is reasonably well supported by such data as have survived on New England plants that sprang up here and there with good or sad results. Thus, while ex-employees of the Undertakers carried their fair share, they can hardly be said to have had a monopoly of the formation of the colonial American iron industry. It is highly probable, however, that the experience they had gained in plants using fine equipment and a sophisticated technology bore good fruit in their work in other and less pretentious surroundings.

If we ask the direct question, "What was the effect of the ironworks of the Company of Undertakers on seventeenth-century Massachusetts?"—the ready answer is, of course, that they met some of the Colony's need for iron. A good quantitative estimate of this need is not available; neither is a sound picture of the ratio of output from Hammersmith and Braintree works to imported iron in filling it. In all likelihood, the latter predominated. It was the competition of iron from the Mother Country which was another of the factors responsible for the relatively early eclipse of New England's first ironworks. Their products, however, were sold, and in good quantity. They were, even judged by today's standards, good products. They can only have been invaluable aids in bringing a comparative wilderness into civilization.

Less obvious but more momentous than either the grinding out of cast and wrought iron or the training of workers was the ironworks' catalytic function. The plans of the founding fathers called for the development of Massachusetts along primarily agricultural and indeed almost feudal lines. A dozen factors threw these plans off. The magistrates and divines envisaged the establishment of a community of true believers committed to the worship and service of God according to a special and narrowly prescribed way. Neither the time nor the place was appropriate. The plans of the Puritan oligarchy, to use Professor Wertenbaker's term,[9] were on the road to failure, however reluctantly abandoned, almost from the time they were put into operation. Commerce and industry had much to do with it. These ironworks, as first ventures in heavy industry, seem to have played a particularly significant role in the total trend.

That the New England ironworks brought the magistrates face to face with special and major problems is clear from the records of the General Court and the inferior judicial bodies. Here were industrial workers of different, and perhaps slight, religious persuasion, rough, cursing, more given to drink than their by no means abstemious neighbors, a very plague on those who sought to lead a people along what they took to be God's way. When they worked they earned the wherewithal with which to get into trouble—if, indeed, wherewithal were ever needed. When they were out of work they had to be cared for. Following the financial disasters of 1653, in particular, these displaced specialists forced the magistrates to wrestle, for the first time in the history of Massachusetts, with the problem of industrial unemployment. At the next higher level of the industrial scale there were also difficulties. With the exception of Winthrop the Younger, and obviously his was a special position, no manager or factor employed by the Undertakers failed, sooner or later, to face the courts on charges of contempt or defiance of magisterial authority. Finally, with the proprietors, there were not only the painful negotiations over special concessions but the task of trying to keep from offending English capitalists who happened also to be highly placed Puritans. One serious lapse in the latter area even led to the magistrates having to answer Oliver Cromwell's objections to their administration of justice.

With little question, too, ironworks problems produced no inconsiderable effect on the laws and judicial system of Massachusetts. In a number of cases it is clear, from date and content, that the impetus to the passage of several rulings and procedural changes came either from ironworks business matters or from snarls developed in the

9 Thomas Jefferson Wertenbaker, *The Puritan Oligarchy: the Founding of American Civilization* (New York, 1947).

almost interminable litigation to which the disasters of 1653 gave rise. In some cases these legislative and judicial findings seem to have served as precedents in the handling of subsequent business problems, although this was not the case in a uniquely fascinating and surprisingly early finding of limited liability of a local shareholder in the Company. Following the Company's bankruptcy, the Massachusetts court calendars were full of suits and countersuits. A full generation of judges and juries wrestled with conflicting claims of ownership, appeals from Massachusetts courts to those of the Mother Country, charges of invalid jurisdiction, and so on almost without end. One who reads the actual records of these legal mazes soon begins to wonder if the Puritan fathers might not have come, with good reason, to rue the day when first they turned a responsive ear to the proposal to erect ironworks in their Colony.

With the ironworks people came a spirit of enterprise. This, obviously, they shared with others in the still young Bay Colony. The workers' indifference to the restraints of the Puritans' social teachings, except when their neighbors were successful in persuading or forcing them to conform, also fostered, in their case, the easier growth of a concomitant spirit of freedom. It can hardly be doubted that both spirits were contagious. The magistrates tried to enforce sumptuary laws, restricting, for example, the wearing of fancy clothing to people of recognized social and economic standing. If the ironworkers managed to get their hands on silver lace, they or their wives were, by their lights, freely entitled to wear them. Much the same essentially non-medieval outlook underlay a workingman's interest in politics, or in religion mixed with politics, which must have been close to zero when the ironworks were first erected, but which by 1661 had a son of Joseph Jenks in court for treason. He had dared to say that "if he hade the King heir he would cutt of his head & mak a football of it."[10] Testimony offered in a number of court cases also made it plain that certain of the workers held no high opinion of local governmental authority, particularly when liquor had loosened lips that ordinarily had closely to be guarded.

Now the role of industry in the colonial period has been little studied. Most of us think of colonial and, indeed, of much of American life as more than anything else an ever mounting push of proprietors of land setting out new plantations and farmers carving out

[10] Mass. Arch., 106/28.

new farms. Commercial activity was also on hand, of course, but in this frontier interpretation of American history, commerce appears largely as a complement of agriculture, the bringer of the things to meet those needs of the farmers which could not be satisfied out of their own exertions. This holds, to a considerable degree, for domestic, intercolonial and transatlantic trade. That farming and lumbering should have had the primacy was axiomatic. The materials on which they thrived were the New World's richest and most readily available natural resources. Colonial records, both southern and northern, make it clear, however, that the early settlers planned, almost from the beginning, to carry on work in mining, metals processing, and manufacturing as well as to farm, fish, and cut timber. The practical results of many and much cherished ventures were disappointing. Nevertheless, industry and the men of industry, entrepreneurs, managers, and workers, were present in America almost surprisingly early and with consequences greater than has ordinarily been appreciated.

Even in later American history the industrial strand has been more than a little shortchanged. It may be that, despite the centrality of industry in making us what we are, we have been the victims of a Jeffersonian interpretation of American history, somehow persuaded that the coming of industry had spoiled the agricultural paradise which our third President took to be God's design for America. Perhaps the blandishments of Turner's frontier interpretation have kept far more than the scholars who have been busily documenting his thesis ever since from seeing that America's growth has rested on commerce and industry as well as on agriculture. Conceivably, the difficulties present in trying to "romanticize" business and manufacturing activities have delayed their acceptance, until quite recently, as proper stuff of history.

In the ironworks erected at Braintree and Lynn, at Hammersmith, the main plant, in particular, we have a case specimen of colonial industry and a forerunner of much that was to come later. In surviving data there is to be found much which is susceptible to romanticizing. For too many people the step from, say, a page copied from the Company's accounts to wholly unsupported stories of trade with pirates has been all too easy. For others a mental picture of Puritans—dressed in clothing which the Puritans would not recognize—toiling in a "quaint" ironworks on the banks of Saugus River is quite irresistible. To tell the ironworks' story in terms of ancestor worship is pointless, if it is not indeed harmful. To tell it otherwise than against an English and New World background and as part of a total economic, technological, and social complex is to do it less than justice.

If it is too much to say that the plants of the Company of Undertakers changed the face of the Massachusetts Bay Colony, it is little to be doubted that they helped to change it. Here were roots not only of much of American iron and steel but of that industrial strand in our history which has so helped Americans to put off the old and take up the new. Hammersmith brought the first effective and sustained production of cast and wrought iron within the limits of what is now the United States. It was a big and bouncing baby, too big for the mother who bore her to sustain and nourish. The baby flourished for but a few years and then died. By our own times only the Ironmaster's House and a tree-covered slag pile were left to mark the scene of what had once been vigorous activity.

While Hammersmith worked, it helped to meet a colony's need for iron. It exerted a hundred influences on its neighbors and the whole community in which it was built. Abandoned and crumbling, it left a legacy. The drives of men which had called it into being did not die. The techniques which men brought here and developed were not forgotten. The changes in patterns of thought and ways of living which it wrought long outlived the stone and wooden structures which industrial pioneers set up along the banks of the Saugus River. Here, indeed, in what it was, what it did, and what it bequeathed, was a worthy prototype of American heavy industry.

2

Prospectors and Precursors

BACK OF THE FIRST successful and sustained reduction of iron ores to useful metal in America, achieved in Massachusetts in the middle of the seventeenth century, were two strands of development, geographically separated but closely interrelated. One was the rise and spread of the English iron industry. The other was the series of efforts of many men over many years to make iron on this side of the Atlantic. The ironworks at Braintree and Lynn were a transplanting of a well-developed technology from a burgeoning English iron trade. They were also the culmination of hopes and dreams of New World explorers and settlers, stimulated in large part by a technological bottleneck arising from the very process of growth which made them possible.

Iron had been made in England from time immemorial. Over the centuries men had worked at simple bloomeries, carrying through from ore to wrought iron in the direct process of iron manufacture. About the time that Columbus made his first voyage, the indirect process, involving the smelting of the ore in the blast furnace and the refining of the iron in forges, was introduced from France.[1] Slowly but surely gathering momentum, the blast furnace took up its vic-

[1] Schubert, "The First English Blast-Furnace," *Journal of the Iron and Steel Institute,* Vol. CLXX (1952), 108–10.

torious march over larger and larger areas. The England from which the later explorers and early settlers came was the scene of what Professor Nef has called an early industrial revolution.[2] The demand for iron had been enormously stepped up. The indirect-process plants made it possible to meet the demand. As the industry spread and as new producing units were added, however, England's forests, the source of the charcoal which ironworks consumed in huge quantities, were increasingly threatened with depletion. One well-known consequence was the series of acts by which Parliament hoped to save the forests and which stand as footnotes to the growth of an industry.[3] Another consequence, less well appreciated today, was an emphasis on iron and other homely metals which looms surprisingly large in the record of English exploration and settlement in the New World.

Now the weighing of motives in that rich complex of factors which drew Europeans to the Americas is no easy matter. In the records of the great voyages and of the early settlements, there are hundreds of references to metals, actual strikes, and ever haunting will-o'-the-wisps. In the South, in the main stream of Spanish activity, gold and silver overpowered all else in motivation and in fact. The Spaniards were not unaware of the presence of iron in their American domains; they were merely too busy to bother with it.[4] Since they got here first and struck it rich, almost beyond belief, it is small wonder that the explorers from rival European countries hoped to do likewise, even though, as late comers, they had to operate in areas where Nature was niggardly of gold and silver. El Dorado cast, and still casts, its fatal spell. There are enough tidings of precious metals just around the corner in the propaganda literature ground out by Hakluyt and Purchas to make one suspect that, while Frenchmen and Englishmen took up the search for iron and copper and the other homelier metals,

[2] John Ulric Nef, "The Progress of Technology and the Growth of Large-Scale Industry in Great Britain, 1540–1640," *Economic History Review*, Vol. V (1934–35), 3–24; "A Comparison of Industrial Growth in France and England from 1540 to 1640," *Journal of Political Economy*, Vol. XLIV (1936), 289–317; and *Industry and Government in France and England, 1540–1640* (Philadelphia, 1940).

[3] Mark Antony Lower, "Historical and Archaeological Notices of the Iron Works of the County of Sussex," *Sussex Archaeological Collections*, II (London, 1849), 190–96; James M. Swank, *History of the Manufacture of Iron in All Ages* (Philadelphia, 1884), 38–39.

[4] Samuel Purchas, *Hakluytus Posthumus, or Purchas His Pilgrims* . . . (reprinted in 20 vols., Glasgow, 1905–1907), XV, 70–71; and XVII, 335.

it was gold and silver that they really wanted. Their interest in baser metals has at least some flavor of a grudging picking up of consolation prizes—when the precious metals were not at hand.

More thorough inspection of the records suggests, however, that the English interest in iron was, at least in some cases, primary and honest rather than secondary and a product of rationalization of the all too cold facts of Nature. Given the picture in the homeland, one of a booming industry choked by a fuel shortage, it would have been strange had it been otherwise. We should expect a difference between Spanish and English attitudes toward metals which stemmed from choice as well as from necessity. We shall see not only that there was such a difference, but that one man, at least, dealt with it in explicit terms. Since France was not experiencing an early industrial revolution, one might also expect to find different emphases among the English and French pioneers in America. Such a pattern is at least faintly discernible. Both were interested in iron. The English got down to real business more quickly and more vigorously than the men of New France.

The story of actual work in iron in the New World began earlier than the arrival of the Spaniards, the French, and the English. It is all but universally agreed that the aborigines of North and South America, unlike many primitive peoples, had failed to master the technique of making iron. A conspicuous exception to this agreement, based on discoveries of supposed prehistoric artifacts and pit- and mound-furnaces, is more interesting than convincing.[5] Some of the natives had iron knives and daggers long before the white man's coming. These, however, came from meteoric iron or, in at least one case, from a telluric iron in Greenland. The iron implements of the Aztecs, which surprised Cortez, and the iron of the Mayas and Incas, valued more highly than gold, were all of meteoric origin and had been worked up cold from stone, so to speak. The Eskimos handled their telluric iron, and meteoric iron as well, in much the same crude way.[6] Normal iron ores were useless to the natives until the introduction of melting at high temperature under blast. This came only with the Europeans and, appropriately enough, with the first Europeans.

[5] A. H. Mallory, *Lost America; the Story of an Iron-Age Civilization Prior to Columbus* (Washington, 1951), esp. chs. 21, 22.
[6] R. J. Forbes, *Metallurgy in Antiquity* (Leiden, 1950), 380, 401–402.

Even as the Norsemen's discovery of the New World antedated
that of Columbus, so they must be given credit for the first manufacture of iron ore in the Americas. According to archeological evidence, it is clear, almost beyond question, that the Norse inhabitants
of Greenland worked up bog iron ores in pot-bowl furnaces and
made iron by the direct process, probably as early as the first colonization of the area. Here an age-old European technique, carried on
in the Scandinavian countries on a domestic and small-scale basis,
largely as a seasonal adjunct to agriculture, migrated from Norway
to Iceland and thence to the Greenland settlements. Iceland seems
to have met its needs for iron over several hundred years. The extent
of production in Greenland is hard to estimate, but the fact that
iron was made there is clear from some physical and much circumstantial evidence. There is nothing to suggest that the Eskimos
adopted the technique. With the extinction of the Norse settlements,
therefore, the making of iron ceased in the northern portions of the
New World. Swank cites a later but fascinating case of the borrowing
of the white man's ironmaking technique by American Indians. In
his time, Cherokees in North Carolina were using Catalan forges, the
appropriate skills having apparently been passed down from generation to generation.[7]

In the records of the heroic exploits of the men of New France,
references to iron, ore deposits, trial processing, and the sending home
of specimens for trial by experts are by no means absent. Cartier,
for example, reporting on his third voyage of 1541, announced that
what we know to be limestone at Cape Rouge was "a goodly Myne
of the best yron in the world" and claimed that the sand on which
they walked was a refined ore all ready to go into a furnace.[8] Early
in the seventeenth century, Champlain found at Sandy Cove, in St.
Mary's Bay, iron ore deposits which his "miner" judged would yield
50 per cent metal. A trial at home proved that he had been wildly
overoptimistic.[9] Lescarbot and Champdore allegedly did something
more wonderful still—finding steel in the rocks from which wedges
and a knife which cut like a razor were made.[10] Concern with iron

---

[7] Swank, *op. cit.*, 75.

[8] *The Voyages of Jacques Cartier* (H. P. Biggar, ed., Ottawa, 1924), 255.

[9] *The Works of Samuel de Champlain* (H. P. Biggar, gen. ed., 6 vols., Toronto,
1922–36), I, 247–48; Marc Lescarbot, *The History of New France* (W. L. Grant, tr.,
3 vols., Toronto, 1907–14), II, 232.

and other minerals, precious and base, is also to be seen at a number of points in the *Jesuit Relations*.[11] On the whole, however, such claims and reports are in marked contrast to the hardheaded and practical approach to iron taken by certain Englishmen, both in Virginia and in Newfoundland.

True, even in the absence of an industrial revolution in the homeland, Frenchmen had reason to keep an eye on the iron potential of New France. Colbert, at least, appreciated the fact that if iron could be made in Canada, it would help to free France of some of the importation from Sweden, on which, as a good mercantilist, he frowned. The men active in real settlement, however, soon reached the conclusion that the riches of Canada were to be found, at least in the early years, in other natural resources. Lescarbot, for example, pointed out in quite realistic fashion that gold and silver were actually deterrents to sound colonization, that iron might be useful, but that there was something far better than any metals until settlement was well established—agriculture.[12] It was, of course, on agriculture, and on quicker and easier extractive operations like fishing and trapping, that the economy of New France came largely to rest. Men took up the really effective reduction of iron ores in Canada, a harder form of extractive work, and one for which we now know Quebec is richly endowed, only comparatively late. Colbert sent de la Potardiere to investigate the iron ore deposits at Baie St. Paul and the bog ores of the St. Maurice valley in 1667. It was only in 1733, however, that the latter were opened up, and not until 1736 that the famous St. Maurice forges near Trois-Rivières were actually making iron from them. These forges, often claimed to be the "first" ironworks in America, are far more significant for their long operating span.

With English explorers in the northern regions the picture was both similar and different. As with the early French travelers, there was no dearth of interest in the precious metals. There was also a counterbalancing concern with baser metals, iron especially, some of it couched in quite realistic terms. True, Frobisher, reporting on his 1576 voyage, pointed out that the natives had iron arrowheads and

---

[10] Purchas, *op. cit.*, XVIII, 281.

[11] *The Jesuit Relations and Allied Documents* (Reuben Gold Thwaites, ed., 73 vols., Cleveland, 1896–1901), III, 67–69; VIII, 13; XXXVIII, 243; LXIII, 289.

[12] Lescarbot, *op. cit.*, II, 235, 317.

knives, but, clearly, he was interested only in gold.[13] Similarly, Captain Edward Haies, in his account of Sir Humphrey Gilbert's expedition of 1583, told of the "minerall man and refiner" turning up iron and silver, and leaving no doubt as to his strong preference for the latter.[14] In the letters of Anthony Parkhurst, on Newfoundland in 1577-78, however, we find not only an interest in iron but stress on the proximity to the ore of rich forests and good water power. He had a thorough appreciation of the threat of iron and saltmaking to the timber of the Mother Country. Nowhere, however, is this emphasis stronger than in William Vaughan's fascinating propaganda piece, *The Golden Fleece*. Vaughan's picture of England is depressing. Her soil was worn out, her wool fallen out of demand, her woods "lately wasted by the Couetousnesse of a few Ironmasters." The latter were Albion's real enemy, the wood shortage the greatest of her problems. The "commodities" of Newfoundland, settlement of which he was urging, were many, but its forests stood out above all else as wherewithal of boards, masts, tanning barks, and charcoal. There is at least a suggestion that he proposed to bring English ore to the timber, although he did say, almost as an afterthought, that "There is great store of Mettals, if they be loo^kt after."[15]

Moving southward and studying the records of men working in the general area of what is now New England, we find, again, an obviously strong concern with metals. Some of the reports are vague but hopeful. Gabriel Archer, reporting on Gosnold's voyage of 1602, told of glittering sands brought up on the sounding lead which might promise "some Minerall matter in the bottome."[16] Similarly Brereton's *A Brief and True Relation of the Northern Part of Virginia*, published in the same year, cited "Stones of a blue metalline colour, which we take to bee Steele oare."[17] Others, however, had a broader perspective and were more restrained in their reporting. John Smith's account of his voyage in 1614 made out a case for settlement based in part on the prospect of cutting down English imports of ship's

[13] Richard Hakluyt, *The Principal Navigations, Voyages, Traffiques & Discoveries of the English Nation* . . . (reprinted in 12 vols., Glasgow, 1903–1905), VII, 227.

[14] *The Voyages and Colonising Enterprises of Sir Humphrey Gilbert* (David Beers Quinn, ed., 2 vols., London, 1940), II, 408.

[15] William Vaughan, *The Golden Fleece Divided into Three Parts* (London, 1626), Pt. I, 4; Pt. III, 13, 15.

[16] Purchas, *op. cit.*, XVIII, 303.

[17] *Ibid.*, XVIII, 321.

stores from Norway and Poland, of iron and rope from Sweden and Russia, and of canvas, wine, steel, iron, and oil from France and Spain.[18] He gave assurance that the Massachusetts coast showed smooth stone for forge and furnace building and iron ore to be worked in them. He also inferred from the resemblance of the coast to that of Devonshire that the cliffs were limestone which would serve as flux.[19] Obviously, he was much in error in all this but, clearly, he had high hopes for iron. On the prospects of gold he disqualified himself, announcing that he was no alchemist, but he flatly stated that anyone who would set up a forge where food and ore and charcoal were free could not lose.[20] Here, too, time was to prove him wrong, partly, at least, because he failed to take into account the by no means small factor of labor costs, even in a wilderness paradise.

Christopher Levett, voyaging along the Maine coast in 1623–24, was far more cautious. He thought the rocks looked like iron stone but added the interesting caveat that if they were not, if would be easy enough to import ore as ship ballast.[21] For most of the early observers, however, there was no question that New England had iron. At the time the Puritans made their settlement, testimony on the point came from many quarters. John White listed iron as one of the things the area could immediately yield, in his *Planters Plea* of 1630.[22] John Smith in his *Advertisements for the Unexperienced Planters of New England or Anywhere,* insisted that it had "Iron Ore none better."[23] Morton's *New English Canaan* told not only of iron in quantity but of several types of ore as well.[24] The author of *A True Relation Concerning the Estate of New-England* excepted iron as he admitted that ore and mineral potentialities were still unknown for

<hr>

[18] John Smith, *Works, 1608–1631* (Edward Arber, ed., 2 vols., Westminster, 1895), II, 711–12.

[19] *Ibid.,* I, 193.

[20] *Ibid.,* I, 201.

[21] *Sailors Narratives of Voyages along the New England Coast* (George Parker Winship, ed., Boston, 1905), 288.

[22] *John White's Planters Plea,* 1630 (reprint, Rockport, Mass., 1930), 26.

[23] Smith, *op. cit.,* II, 927.

[24] Thomas Morton, *New English Canaan; or New Canaan Containing an Abstract of New England* (Charles Green, pr., 1632), 58 (in Peter Force, *Tracts and Other Papers, Relating Principally to the Origin, Settlement and Progress of the Colonies in North America* [4 vols., 1836–46], II). (Hereinafter referred to as Force, *Tracts.*)

want of trial.[25] Finally, William Wood, who had come over in 1629, commissioned to deliver a full report on the resources and general character of the territory for the information of English Puritan leaders, announced in his *New Englands Prospect* that he had had definite word of iron stone and of black lead, had seen what appeared to be lead ore, and had heard from the Indians that gold, the "Spaniards blisse," might lie hidden in the barren mountains.[26] Interestingly enough, two of the minerals mentioned, the iron and the graphite, were in time to be exploited by John Winthrop the Younger, the son of the leader of the Massachusetts Bay Company's settlements in the New World.

It was, however, in the region still farther to the south that major English settlement was first accomplished. There we find, and can better analyze, because of far better surveying data, a concern with mineral prospects that was vigorous from initial explorations down to, and for many years following, the well-known iron venture at Falling Creek, Virginia, which antedated even the settlement of Boston. In the earliest activity, that of the Roanoke Island settlers under the sponsorship of Sir Walter Raleigh and the leadership of Ralph Lane, it was gold on which men's eyes were fixed. Nevertheless, Thomas Hariot, historian of the short-lived colony, duly recorded various mineral possibilities, including those of iron and alum. Of the former he said:

In two places of the countrey specially, one about fourescore, & the other six score miles from the fort or place where we dwelt, we found nere the water side the ground to be rocky, which by the triall of a Minerall man was found to holde iron richly. It is found in many places of the country els: I know nothing to the contrary, but that it may be allowed for a good merchantable commodity, considering there the small charge for the labour & feeding of men, the infinite store of wood, the want of wood & deerenesse thereof in England and the necessity of ballasting of ships.[27]

This is the first discovery of iron ore within the limits of the United States. Dating from 1585, and described in Hariot's *A Report on Virginia* of 1587, it was not exploited, and ironmaking in North Caro-

[25] Transcription of Ms. in British Museum by Henry F. Waters, with notes by Charles E. Banks (Boston, 1886), 9.
[26] London, 1634 (reprint, Boston, 1865), 15–16.
[27] Hakluyt, *op. cit.*, VIII, 356.

lina was not achieved until long after successful efforts in several other colonies.

Priority in terms of exportation of true iron ores, as, indeed, of successful English colonization, goes to Virginia. John Smith, always on the lookout for minerals, sent to the Treasurer and Council of Virginia what he described in a letter dated 1608 as "two barrels of stones, and such as I take to be good Iron ore at the least."[28] In the same year the Company's ship, Christopher Newport, captain, reached England with freight of sassafras, cedar, walnut, and iron ore. The latter was smelted down and from it were made several tons of iron which were sold to the East India Company with glowing endorsements as to quality.[29] This was more than sufficient to give iron a key role in the subsequent surveying and description of Virginia's mineral resources.

While high hopes for finding gold were slow to die, and there are many references in the early literature to "hills and mountains making a sensible proffer of hidden treasure, neuer yet searched,"[30] and the like, the story of minerals in Virginia is richer in caution and realism than in quest of will-o-the-wisps. Gold might be found but iron was right at hand. Sir Thomas Gates assured the Council in 1610 that iron ore lay on the ground over a ten-mile circuit.[31] William Strachey's *Historie of Travaile in Virginia* reported on a "goodlie iron myne" from which Newport's shipment had come, announcing that the East India merchants preferred it to that of any country.[32] Alexander Whitaker's *Good Newes from Virginia, 1613* glowingly claimed that iron and steel, among other things, had "rather offered themselves to our eyes and hands, than bin sought for of vs."[33]

Here, too, are to be found statements relating the prospects of ironmaking in Virginia to the wood shortage back home. Take, for example, the *Nova Britannia: Offering Most Excellent fruites by*

[28] *The Genesis of the United States* (Alexander Brown, ed., 2 vols., Boston and New York, 1891), I, 203. (Hereinafter referred to as Brown, *Genesis.)*

[29] William Strachey, *The Historie of Travaile in Virginia Brittania* (R. H. Major, ed., London, 1849), 132.

[30] *Nova Brittania: Offering Most excellent fruites by Planting in Virginia* (London, 1609), 11 (Force, *Tracts*, I).

[31] Purchas, *op cit.*, XIX, 71–72.

[32] Strachey, *op. cit.*, 131–32.

[33] Reprint (New York, 1936), 39.

Planting in Virginia, published in London in 1609. Its author offered
a quite realistic situation estimate:

And from thence (Virginia) we may haue Iron and Copper also in great
quantitie, about which the expence and waste of woode, as also for building
of Shippes, will be no hurt, but great seruice to that countrey; the great
superfluity whereof, the continuall cutting downe, in many hundred yeares,
will not be able to ouercome, whereby will likewise grow a greater benefite
to this land, in preseruing our woodes and tymber at home, so infinitely and
without measure, vpon these occasions cutte downe, and falne to such a
sicknesse and wasting consumption, as all the physick in England cannot
cure.[34]

Similarly, the Council of Virginia in *A True Declaration of the estate
of the Colonie in Virginia* of the following year stated that there only
men's labor was lacking to furnish the homeland with timber which
the ironworks and excessive building had consumed to such a degree
that the Navy's needs a few years hence could not be met by either
the "scattered Forrests of England" or the "diminished Groues
of Ireland."[35]

The ores had been tested. Men were sure that, in her stands of
timber, Virginia offered a cure for England's dwindling wood supply.
The need for iron in the colony itself was self-evident. Nature's gifts
and common sense dictated that the reduction of iron ore should be
one of the first industrial ventures to which the Virginia colonists
should turn. Amazingly enough, efforts in this direction were planned,
and faltering first steps taken, as early as 1609–10, probably as soon
as word came back of the good promise of the ore carried to England
by Captain Newport.

Data on these first attempts to make iron ore are fragmentary.
In an appendix to *A True and Sincere declaration for the purpose
and ends of the Plantation begun in Virginia,* entered at Stationers
Hall in December, 1609, is a list of artisans required for Virginia.
Among them are ten iron men for the furnace and two "minerall
men."[36] The same are also listed in *A Publication by the Counsell of
Virginia touching the Plantation There,* of 1610.[37] A more impressive

---

[34] *Nova Brittania* . . . , 16.
[35] London, 1610, p. 25 (Force, *Tracts,* III).
[36] Brown, *Genesis,* I, 353.
[37] *Ibid.,* I, 356.

enumeration, including two millwrights for iron mills, four iron miners, two iron founders, two hammermen for iron, edge-tool makers for ironwork, two colliers and two woodcutters, appears in *The Tradesmen to be sent into Virginia under the Comaunde of Sir Thomas Gates*.[38] This may all have been a matter of paper proposals, but there is some reason to believe that, in this first phase of Virginia ironmaking, more than plans and projects was involved.

The instructions to Gates in May, 1609, suggested that he seek the advice of, among others, "Captaine John Martine (and) Captaine Richard Waldoe, M^r of the workes."[39] It is not clear what works was meant, and we have no data on Waldoe's professional background. In June, 1610, however, Captain John Martin was nominated "Master of the B(attery) Workes for steele and iron"[40]—a rather tight designation for works still wholly visionary. Martin was not an iron expert by any means. It is possible that he was taking on responsibility for ironmaking, perhaps on an additional duty basis, after Waldoe's death. If this supposition is correct, the obscure Waldoe would qualify as the first "specialist" charged with the work of making Virginia's much-heralded iron resources pay off.

Of the location and equipment involved in this initial venture we know nothing. We are not informed of the origins, numbers, or types of workers in iron who may have been employed. If we accept the judgment of Sir George Yeardley expressed in a letter to Sir Henry Peyton in November, 1610, we might conclude that the successful reduction of ore to metal was still in the future, waiting on adequate capital and the right men.[41] It could be argued, however, that some iron was made at this time, probably on a test-sample basis. First, Francis Maguel, an Irishman in the service of Spain, reported to Madrid in 1610 that in Virginia there were "so many iron mines (to work which, as well as to work other metals they have already erected there some machinery) it will be very easy to them to build many ships."[42] Next, when many years later the Company was defending itself against criticism of its policies, it claimed that ore of good

38 *Ibid.*, I, 469.
39 *The Records of the Virginia Company of London* (Susan Myra Kingsbury, ed., 4 vols., Washington, 1906–35), III, 13. (Hereinafter referred to as *Va. Records.)*
40 Brown, *Genesis,* I, 408.
41 *Va. Records,* III, 30.
42 Brown, *Genesis,* I, 398.

quality had been found in abundance and that it had been brought over "wrought and unwrought into this Kingdome."[43] The context of this vague statement suggests that this importation of iron, and *wrought* iron ore must be iron, was in this general period. If, however, really impressive results had been obtained, the Company, here and elsewhere, would certainly have publicized them. Apparently, therefore, whether from the exodus of workers to tobacco cultivation or otherwise, the real job was still to be accomplished.

It was, in fact, about another decade before ironmaking was again actively pressed. Back of these new efforts were some rather strange developments in the Mother Country. In 1619 an anonymous philanthropist who signed himself "Dust and Ashes" deposited £550 with the Virginia Company for "the bringinge vp of the Infidells Children in true religion and christianity."[44] In due time the Company decided to entrust the money to two subsidiary groups, the men of Southampton and Martin's Hundred, to be spent on bringing up a number of children in the Christian faith and a useful trade. The adventurers of Martin's Hundred were able to duck an annoying chore by pleading the confused state of their Virginia property. The Southampton people tried to do likewise but failed, though they offered to add £100 to the original donation on condition that the responsibility "might not be put vpon them." On the Company's insistence, they finally accepted the burden—but saw fit to reach the donor's aims by considerable indirection.

Present-day Americans are accustomed to business giving birth to philanthropy. Here a charitable contribution provided the impetus, and at least a part of the wherewithal, for America's first major ironmaking effort. After careful and, one infers, somewhat strained consideration "howe this great and waightie buissines might with most speed and great aduantag be effected, . . . it was agreed and resolued by them to imploy the said money together with an Addicon (out of the Societies purse) of a farr greater Some toward the furnishinge out of Captaine Bluett and his Companie being 80 verie able and

---

[43] *Va. Records*, IV, 141. See also, *ibid.*, I, 472. It is of course possible that Sandys was here referring to iron ores, perhaps to the Newport shipment, in his reference to "Iron sent from thence."

[44] *Ibid.*, I, 587. "Dust and Ashes" was identified in testimony before an October, 1624 court as a "Mr. Barber" (*Minutes of the Council and General Court of Virginia, 1622–1632, 1670–1676* [H. R. McIlwaine, ed., Richmond, 1924,] 24.)

sufficient workmen wth all manner of prouisions for the setting vp of an iron worke in Virginia, whereof the proffitte accruinge were intended and ordered in a ratable proporcon to be faithfully imployed for the educatinge of 30 of the Infidelle Children in Christian Religion and otherwise as the Donor had required." In commending the project to Yeardley, who was both governor of Virginia and captain of Southampton Hundred, it was described as "a worke whereon the eyes of God, Angells, and men were fixed."[45] Few American business enterprises can claim to have begun under such auspices!

The bulk of the Bluett ironmaking venture is clouded in mystery. To judge from the number of workmen, the project was conceived in the grand manner. Of fiscal and supply arrangements we know only that they were similar to those agreed upon with a successor, the key figure in Virginia ironmaking, to whom we shall shortly come. It is not clear whether it was intended that Bluett erect a plant at the scene of the 1609–10 activities, or at a new one. Certainly, he got to America; equally certainly, he became ill and died shortly after his arrival.[46] What had been accomplished in the interval, and what finally happened to his workmen, are quite obscure.[47] It is doubtful if real progress had been made. For all we can tell from existing data, this venture of the Southampton Hundred, with the philanthropies of "Dust and Ashes" lurking in the background, must be pronounced a failure.

Impressive as the scale of Bluett's project seems to have been, his was but one of three ironmaking ventures in the 1619–20 period. Of the number there can be no question. The Company Records for 1619 tell us specifically that some one hundred and fifty men had been sent to Virginia "to set up three Iron workes."[48] All were apparently specialists, since they are indicated in another document as having come in the number of "about one hundred and ten" from Warwickshire and Staffordshire, and of "about forty" from Sussex, "all framed to *Iron*-workes."[49] Three "master workmen" went over

---

[45] *Va. Records,* I, 587–88.

[46] *Ibid.,* I, 588.

[47] *Ibid.,* III, 128, 446–47. The survivors were probably drawn off to agriculture, the cultivation of "this vicious weed of Tobacco." (Sir William Berkeley, *A Discourse and View of Virginia* [London, 1633]).

[48] *Va. Records,* III, 116.

[49] *Ibid.,* III, 309.

—and promptly died.[50] They were replaced. At least we learn from a note on the shipping, men and provisions sent to Virginia in 1620, that "There are three principall men sent againe for Masters of the Iron *works* which are in some good forwardness, and a proof is sent of Iron there made."[51] The last items suggest that real progress had been made. So, too, we might conclude from a letter to Sir Edwin Sandys dated July, 1621, which reported that "the Iron workes goeth forward veary well."[52]

Our data are so thin and contradictory, however, that it is all but impossible to pin down details of sponsorship, management, working staff, and plant locations, let alone settle the question of which, if any, actually turned out iron. Of the three ventures, two were under the aegis of the Southampton adventurers. All, by one estimate, were failures, in each case because their people "were not able to mannage an Iron worke and soe turned good honest Tobaccoe mongers."[53] Of the six "master workmen," the original trio and their replacements, we can identify only Bluett. The working crews of all three were doubtless included in the rough total of one hundred and fifty mentioned above. Bluett's group consisted of eighty men. Another, apparently not under Southampton auspices but somehow connected with the project to convert and educate Indians, seems to have had fifty. This would suggest that a second project of the Southampton group, one on which no data at all have survived, must have been staffed by about twenty. Assuming these were all specialists, it looks as though Virginia had had enough people on hand to make iron.

In all the literature there is only one reference in which doubt that so many ironworkers were sent to the colony got expressed.[54] And, on the other hand, there were at one point plans for still another shipment of workers. In mid-March, 1620, the Treasurer of the Company reported on "one m^r King that is to goe wth 50 persons wth him to Virginia there to sett on foote Iron Workes."[55] It is almost certain that this group did not sail. The negotiations were taking place after the one hundred and fifty had reached Virginia; had they

[50] *Ibid.*, I, 472.
[51] *Ibid.*, III, 240.
[52] *Ibid.*, III, 464.
[53] *Ibid.*, IV, 141.
[54] *Ibid.*, IV, 185.
[55] *Ibid.*, I, 322.

gone, the colony would presumably have had some two hundred specialists in the making of iron. This, one is tempted to say, would have been close to overpopulation. Surely, however, no one can doubt that the Company was enormously interested in making the Virginia ores pay off. Three ironworks in a struggling colony seem actually to have been attempted. All, so far as we can tell, were failures. And by May of 1621 the quite impressive sum of £4000, according to one estimate, had been spent on ironmaking in Virginia.[56]

In May, 1621, Sir Edwin Sandys informed the Company that he had received "credible informacon" that three of their master workmen were dead, that their replacements had been sent over, and "now it was their good happ to light vpon a fourth gent (named m^r Iohn Berkly who in the iudgement of those that knowe him well was helld to be very sufficient that waye) who did now offer himsellf to goe vpon the said service and carry over wth him 20 principall workemen well experienced in those kinde of works."[57] Thus began what seems to have been wholly an ironmaking venture of the parent company. The pious intentions of "Dust and Ashes" were not forgotten; indeed, it was assumed that success here would mean that at too long last they might be realized.[58] But the Southampton group, whether from the dissatisfaction of the philanthropist with earlier arrangements, or from sensitivities about further draining of its "private purses," or from a conviction that, if Virginia were ever to have an ironworks, it would take a major effort by the Company proper, was here called on only to provide certain supplies to Berkeley's group when it got to the colony.[59] All the negotiations with this fourth iron expert were carried on by the court and committees of the Virginia Company.

The conditions of Berkeley's hiring were like those for Bluett. He was to get £20 for procuring the workmen and getting them to the Isle of Wight in time to sail, and £30 to cover his own provisioning, the transportation, furnishing and feeding of himself, his son, three servants, and the workmen, for a full year. He and his son, Maurice, agreed, on their side, to serve, along with the workmen, for seven years. The latters' prospective functions leave no doubt as to the scale of enterprise anticipated. Eight were to work at the blast

[56] *Ibid.*, I, 472.
[57] *Idem.*
[58] *Ibid.*, I, 588.
[59] *Ibid.*, I, 586; III, 651; IV, 141.

furnace—two as founders, keepers, fillers, and carpenters, respectively. Twelve were forge men—four finers and their two assistants, two chafery men, two hammermen, and their two helpers. The arrangements proving satisfactory, Berkeley and these men, and apparently he was able to recruit the full number in a tight labor market, set sail for Virginia, presumably in July, 1621.[60]

The terms in which the Berkeley party's coming got announced to the people in Virginia suggest that this was to be a last-ditch stand in the trying business of extracting iron from the by now so-long-publicized ores. Ironworkers already on hand were to be put in "some secondarie or subordynarie places of assistance to m^r Berkly," should their own efforts not succeed "by want of workes or necessarie Materialls." The Governor and Council were asked to provide good entertainment at the newcomers' landing, and all possible assistance then and thereafter. Then, as though appeal were not enough, the same authorities were asked to extend their authority to the utmost to force any recalcitrants to come through with aid to "this worke of so geat Consequence & generall expectacon infinitt Com(m)°(dity) & vnspeakable Benifitt."[61]

When the new group reached Virginia is not known. Neither may we be certain that the Berkeleys concentrated their building efforts at a site selected by Bluett or one of the other "master workmen." That something must have been wrong with the earlier locations may be inferred from another letter to Sandys, this one dated June 12, 1620, which suggested that the whole business should have been more carefully deliberated in England, and experts should have spent a whole year seeking out abundant ore and good sites for ironworks building.[62] On the other hand, Berkeley did not spend a year in location surveying. Indeed, letters of his reaching England late in 1621 or early in 1622 indicated not only that he found Virginia a fit place for ironworks "both for wood, water, myne and stone," but that he was confident of actually making iron by the following Whitsuntide.[63] Unless Berkeley had been at least a little overoptimistic, and it should be noted that he was asking for additional funds

[60] *Ibid.,* I, 475–76; III, 475–76.

[61] *Ibid.,* III, 476, 487.

[62] *Ibid.,* III, 303.

[63] Edward Waterhouse, *A Declaration of the State of the Colonie and Affaires in Virginia* (London, 1622) and in a broadside issued in May, 1622. *(Ibid.,* III, 548, 640.)

as he passed along the bright tidings, one would have to conclude from the brief time span involved that he had chosen to pick up where one of his predecessors had quit.

Whether new or old, the site in question was on the banks of Falling Creek, not far from the present city of Richmond. The location was described by Sandys in a letter to the Company as so suited to the manufacture of iron—

as if Nature itself had applyed her selfe to the wish and direction of the Workeman; where also were great stones hardly seene else where in Virginia, lying on the place, as though they had beene brought thither to aduance the erection of those Workes.[64]

It is difficult to tell to what extent Berkeley and his people had brought things at this spot, however well favored by Nature, however well supported his efforts by the settled colonists, when he and his workers got cut down in the Indian massacre of March 22, 1622. According to one source, twenty-seven persons were killed; there were, presumably, but two survivors, a boy and a girl.[65] Maurice Berkeley was not at Falling Creek at the time and also survived. All or nearly all of the effective working crew undoubtedly must have perished; the ironworks was ruined, and the workmen's tools thrown into the river.[66]

It has generally been assumed that the Indians destroyed a plant which was well along to completion. Indeed, according to one authority, slag from the Falling Creek furnace had been found in the nineteenth century and was, in the first decade of the present century, preserved in the Virginia Historical Society.[67] If the material was

[64] Waterhouse, *op. cit.*, (*Ibid.*, III, 548).

[65] Waterhouse, *op. cit.*, (*Ibid.*, III, 565). Survivors are mentioned in Robert Beverley, *The History and Present State of Virginia* (London, 1705) (Louis B. Wright, ed., Chapel Hill, 1947), 55.

[66] S. P. Col., IV, fol. 45 (Sainsbury Abstracts for 1628 [Va. State Library], 178) (summarized, *Calendar of State Papers, Colonial Series* [London, 1860–19], America and West Indies, 1574–1660, 89); Conway Robinson, "Notes from the Council and General Court Records," *The Virginia Magazine of History and Biography*, Vol. XIII (1906), 397.

[67] Philip Alexander Bruce, *Economic History of Virginia in the Seventeenth Century: an Inquiry into the Material Condition of the People, Based upon Original and Contemporaneous Records* (2 vols., New York and London, 1907), II, 449, n. 2. The slag specimens seem to have been lost in the intervening years. It is conceivable that it was the product of eighteenth-century forge operations in the same general area.

in fact furnace slag, the furnace must have been finished and in blast. But if the building of a blast furnace took time, its destruction by savages would have been no easy matter. Unless there had been quite complete devastation, one might expect that repairs and the replacement of workers and working tools would have enabled others to pick up the pieces and get into production. They did not. The disaster of Falling Creek put an absolute quietus on activities there and on ironmaking in Virginia for many years. Making all allowance for deficiencies of capital and lack of skilled workmen, one must wonder if the ironworks had been as close to a going concern as Virginia historians have suggested.

It took time, of course, for news of the disaster to reach England. In the interim, Berkeley's good news had been received and the Company persuaded to advance additional capital, presumably by special assessment on the shareholders.[68] This was in April. In June, the Company was still cherishing high hopes. It wrote the Virginia authorities, asking that "some good quantitie of Iron and Myne" be included in the next returns.[69] After the catastrophe the Company was still insisting that success in ironmaking had been just around the corner when the Indians let loose their attack, claiming now that the ironworks had been brought "in a manner to pfeccon," now that it had been "in a very great forwardnes."[70]

There were, however, evidences on the "other" side. In Beverley's *History and Present State of Virginia* we again see that the plant had been nearly finished and learn that the ore had proven "reasonably good." On the other hand, we are also told that "before they got into the body of the Mine, the people were cut off in that fatal Massacre."[71] Similarly, when in August, 1622, the Company, by then informed of the tragedy, wrote to have any surviving ironworkers placed in Maurice Berkeley's charge, and to ask for a list of their names and "professions" and a "noat of the tools and materialls wanting for the executing of the worke," the general tone of the letter hardly suggests a quick picking up from an all but going ironworks. Indeed, the authorities stated that the making of iron, however vital, would have to wait until such a time as they could "againe

68 *Va. Records,* I, 629.
69 *Ibid.,* III, 646.
70 *Ibid.,* II, 349, 384.
71 Beverley, *op. cit.,* 49, 126.

renue that bussiness so many times vnfortunately attempted," quite as though they were writing off a fourth ironmaking project as one more failure.[72]

The tragedy at Falling Creek seems to have been the last straw to the critics of the Sandys administration, which had sought not only to erect ironworks but to promote other kinds of industrial activity and develop new forms of agriculture. Sir Nathaniel Rich, for example, in a "Draft of Instructions to the Commissioners to investigate Virginia Affairs" in April, 1623, wrote that "The vast and wilde proiectes of Sir E. S. have ruyned plantacons," and brought up as evidence "4 Iron workes at once all of them by halfes where as one or 2 might to be effected."[73] The man responsible for "Parts of a Draft of a Statement touching the Miserable Condition of Virginia" wrote, early in the summer of the same year, of ". . . many wilde & vaste pjects set on foot all at one time," including "3 Iron works," and pointed out that all had come to naught because they were all "inioyned to be effected in the space of 2 years, by a handfull of men that were not able to build houses, plant corne to lodge and feed themselves . . . ."[74]

The rebuttal to such charges showed courage—and ingenuity. As early as Berkeley's departure, the Sandys group was justifying the multiplicity of projected ironworks on a basis of sober caution. If some failed, others might come through. Now it summoned up vague attacks on its attackers, pointing out, in one instance, that the troubles had been due in part, at least, to "emulous and envious reports of ill willers whose pryvate ends by time wilbe discouered and by God recompenced." Again, and often, it insisted that only the massacre had stood in the way of real accomplishment. This defense, however, was neatly turned by one of the critics who insisted that the disaster had been but "the fayrest excuse for all errors and might haue served the relato[rs] turne very well and made them bould to affirme that worke was wholly over throwne by it. . . ." In the welter of claims and counter claims, two things stand out. One was the bill for four years' effort to make iron—£5000. The other was the return on the investment—"a fire shovell and tongs and a little barre of Iron made by a Blomery."[75]

[72] *Va. Records*, III, 670–71.
[73] *Ibid.*, IV, 117.
[74] *Ibid.*, IV, 176–78.
[75] *Ibid.*, III, 384, 476; IV, 141.

Since this report came from critics of the then dominant group, we must conclude that some iron was made in Virginia in the second phase of operations. It is not at all certain, however, that the items came from the plant at Falling Creek. All of them were doubtless made of wrought iron. All of them could have come from bloomery or direct-process operations. Berkeley's plans, by inference from the breakdown of his workers, called for ironmaking by the combination of blast furnace and forge. These products may have been made by Berkeley in a crude direct-process forge, perhaps in order to impress his sponsors back home. They may have come from one of the other ironworks, even though the bulk of surviving documentary evidence suggests that these had been abortive ventures. Neither our data nor presently available physical evidence at the Falling Creek site permit us to estimate the degree of success achieved by Berkeley before the Indians' blow struck.[76] But there is little reason to doubt that, whether because of the massacre or other and more deeply rooted factors, Berkeley's work was no more effective in getting Virginia ores made into iron than that of his predecessors.

As to other factors, we have only a few clues. The Council in Virginia wrote the parent company in January, 1624, that it wholly lacked the means to re-establish the plant at Falling Creek. In the meanwhile, the Company had been doing some pondering of its own and had indicated, in a letter dated August, 1623, that it proposed once again to make iron. This time, however, it was a bloomery which the authorities had in mind, and one whose costs would be borne by a few private adventurers. This was explicitly contrasted with the "greater waies wch we haue formerly attempted." Hope was also expressed that success of the more modest plant would be a "great inducement and ground for the maine worke themselues."[77] We cannot tell, of course, whether the choice here of operations of quite limited scope, and of what seems to be profit-motivated capital risk, hinged on belated awareness of the colony's economic scale and dynamics, or on sheer necessity. Of the project itself, we know only

---

[76] Thorough search of the Falling Creek area by the archeologist of the Saugus Ironworks Restoration and the author in May, 1951, disclosed no traces of furnace activity, no sites that appeared to meet the generally prevailing criteria for seventeenth-century blast-furnace building. There has been considerable topographic change in the general vicinity in recent years.

[77] *Va. Records,* IV, 268, 452.

that the adventurers preferred to have the works set up in Martin's Hundred, and left the naming of a director to the people of the colony. In all probability, it was not actually carried through. But it made more sense than the grandiose schemes attempted earlier.

Interestingly enough, it was only after all these unfortunate ventures in the field of ironmaking that the world could read the opinion of a well-informed and eloquent man who openly weighed the Spanish experience with precious metals and the English hopes for Virginia iron in the balance—and registered a potent preference for the latter. This was the author of *Virginias Verger*, published in 1625 in *Hakluytus Posthumus, or Purchas his Pilgrimes*. Were gold and silver the only source of wealth, he asked. The minds of those who looked contemptuously on Virginia because, so far, only iron had been found were, he insisted, of a "sordid tincture" and a "base alloy." After all, had not the Spanish iron drawn to it the Indians' silver and gold? Had not "English Iron brought home the Spanish-Indian Silver and Gold?"[78] But one of several statements testifying to the importance of iron in the eyes of English explorers and settlers, as we saw earlier, it is unique in its setting off of the precious and base metals, one against the other. Unfortunately, the judgment of this man who spoke for an "Iron age of the world," as well as for Virginia, was confounded by actual experience in the colony, both before and long after it was put in print.

The rest of the story of ironmaking in Virginia is a matter of further proposals and projects, none of which came through. One, in 1628, called for the incorporation of a number of Englishmen as an iron-producing company with a fourteen-year monopoly, subsidies, customs exemptions, etc. We have no record that the group was chartered. In the same year ore was again sent to England, the local authorities pointing out that the establishment of ironworks was quite beyond their means. What came of this is wholly obscure. Two years later, Governor Harvey went to Falling Creek to determine if further work at the site were feasible. He wrote the English authorities that it met all necessary conditions with respect to water power, timber, ore, and building stone. Significantly, he did not comment on existing ironworks remains. This might suggest either that the destructive work of the Indians had been so thorough as to be almost incredible

[78] Purchas, *op. cit.*, XIX, 232–33.

or that work had not been far along at the time of the massacre. There seems to have been at least one local and private ironmaking venture, that of Sir John Zouch and his son, of which we know only that it failed for lack of support from some unidentified partners.[79] Even the expressed willingness of the King to erect ironworks at his own expense came to naught, perhaps in consequence of a more or less official report that there was not enough ore in Virginia to keep one plant going for seven years.[80]

Despite this judgment there was still no end to the propaganda for work in iron in Virginia. Men were as hopeful as the first explorers, but the long record of grim experience had at least taught that the step from iron ore to useful metal is not easily taken in a colony being carved out of the wilderness. Something special was needed. The author of *A Perfect Description of Virginia* drew as fine a picture of Virginia's promise in iron as we have seen. The ore was good. The timber supply was inexhaustible. There were fine sites, convenient to building stones, water power, and water transportation. Not only could an ironworks meet the colony's needs, but it could provide muskets and armor and other things for trade. It could become "the Magazine of Iron Instruments in every kinde, and at cheape rates; so that no Nation could afford them halfe so cheap. . . ." All this could be had in six months' time, given twenty imported specialists, the right tools, and adequate supplies, if—and this is important as perhaps the earliest proposal of employee profit sharing in America —the workers were given half of the annual net return on operations![81] Similarly, John Clayton, in a letter to the Royal Society in 1688, glowingly summarized the natural advantages. He offered no special solution, but he was aware that the making of iron would require more capital than the colonists could afford, and further pointed out that "for Persons in *England* to meddle therewith, is certainly to be cheated at such a Distance."[82]

By this time, indeed by the time the profit-sharing scheme was

---

[79] Bruce, *op. cit.*, II, 451. Since Harvey was making out a case for the Falling Creek site's high promise, one would expect him to have mentioned an even partially standing furnace stack, the repair of which would certainly have cost less than the erection of a new furnace.

[80] *Ibid.*, II, 453.

[81] *A Perfect Description of Virginia* (1640), 4–5 (Force, *Tracts*, II).

[82] *A Letter from Mr. John Clayton to the Royal Society*, 27 (Force, *Tracts*, III).

proposed, ironworks had been successfully established in Massachusetts. At the Braintree and Lynn plants of the Company of Undertakers of the Iron Works in New England, there was no employee profit sharing. Here, however, as we shall see, the exploitation of local ores took the importation of skilled workmen, special support from "government" of the type called for in the Virginia proposals of 1628, and a huge capital investment, one far beyond the reach of Massachusetts residents. Interestingly enough, the men who put up the money came to feel that they had been cheated, and the distance which separated them from their investment played a major role in the business. The story of these plants abounds in obstacles, difficulties, and tribulations.

Taken as a whole, the failures of the various efforts to process iron ore in Virginia, which we have recorded here, must be attributed to more deeply rooted factors than hostile Indians. The right combination of men, materials, money, and market did not occur. Something like the right one did appear in Massachusetts. As we turn to its story, however, it is fair to point out that if we call the Virginia ironmaking projects "failures" and those of the Puritan settlement to the north, "successes," the line between them is narrow indeed. They had much in common, almost everything but hostile Indians. It may well be that, regardless of the degree of completion to which Berkeley had brought his plant by the fateful day of March 22, 1622, only the tragic mishap of the massacre in Virginia left Massachusetts in the position to claim the palm for the first manufacture of iron on an industrial basis within the limits of what is now the United States.

3

John Winthrop
the Younger, Promoter

MANY AN EXPLORING VOYAGE had been motivated by the lure of metals. Many a traveler had returned home with wondrous tales of America's mineral wealth—tales to be taken on faith, or borne out, or dashed forever by analysis of samples taken along as evidence. When the Massachusetts Bay Company launched its large-scale settlement, it was axiomatic that the possibilities of mineral exploitation should loom large. Its people had grown up on the propaganda literature. Interested in trade and industry, they had an eager eye for the natural resources, particularly the mineral resources, which were in many respects basic to both. The prospective Puritan emigrants differed from many of their predecessors, however, in the systematic plan making which underlay their venture in general, and in the soberly realistic attitude with which they approached the prospects of metallic resources in particular. Theirs was to be no quest for a will-o-the-wisp if they could help it.

The early records of the Company are full of data on the outfitting of vessels bound for New England, and bear ample witness to the range and calibre of the Puritans' planning. There is no reason here to describe the logistics of the migration. Even a casual scrutiny of the *Records* and the *Winthrop Papers* will show the care with which all kinds of supplies were ordered and assembled for the transatlantic voyage.[1] In addition to taking on food, clothing, medicine, tools, and all the things needed to carve out a new settlement in the wilderness, the Company tried to recruit, systematically, key specialists whose skills might one day lead to the development of various industrial ventures and make the new colony less dependent on importation from England. Among these, experts in metals are quite conspicuous.

Under date of March 2, 1629, the Company recorded, "Also, for Mr. Malbon, it was propounded, he having skill in iron works and willing to put in £25 in stock, it should be accepted as £50, and his charges to be borne out and home for New England; and upon his return, and report what may be done about iron works, consideration to be had of proceeding therein accordingly, and further recompence, if there be cause to entertain him." Whether because Malbon had difficulty in making up his mind or because the Company was bent on making assurance doubly sure, within three days it was also considering the qualifications of one Thomas Graves for the mineral development job. He had a broader range of capacities, being "a man experienced in iron works, in salt works, in measuring and surveying of lands, and in fortifications, in lead, copper, and alum mines, etc."[2]

On March 10, Graves entered into formal agreement with the Company to go to New England at its charge, with assurance of transportation for a wife, five children, and two servants, and assurance of housing, subsistence, and a land grant. Remuneration was set at £15 per month on a short-term basis and at £50 per year should he choose to remain in the Company's employ.[3] The Company regarded this as a major investment, as is clear in a letter to Captain Endicott at Salem in which the latter was urged to make careful trial of his "sufficiency." After reciting his qualifications and urging that his advice be followed in the location of settlements, the Company pointed out, "His salary costs this Company a great sum of money, besides which (if he remains with us) the transporting of his wife, and building him a house, will be very chargeable, which we pray you take into your consideration, that so we may continue or surcease this charge as occasion shall require."[4]

[1] *Winthrop Papers,* II, 171, 177, 215–18, 247–78; *Mass. Records,* I, 23–27, 31–32, 35–37.
[2] *Mass. Records,* I, 28, 30.
[3] *Ibid,* I, 33.
[4] *Ibid.,* I, 390–91.

Both experts came to New England, reaching Salem with Higginson's fleet, in 1629. Malbon is believed to have returned home the following year, but neither the fact nor the reasons therefor are clear. Graves stayed longer, becoming a freeman in the spring of 1631, but probably went back to England in 1632 or 1633.[5] Data on the gentlemen's metallurgical activities have not survived, and it is impossible to tell if, after practical trial of their abilities, the Company decided it could dispense with their services. Graves would seem to rate some kind of recognition as a "first" American engineer, but obviously, his claims rest on the thinnest of documentary data.

With the return of the specialists to the homeland, prospecting for minerals became wholly incidental to the general activities of settlement. There is record of only one further attempt to do something with iron in the first decade of Massachusetts history. Of this we know little more than that death brought it to a halt, but, probably, the man involved was a regular settler with an eye out for metals, rather than a specially imported expert. The General Court of November 2, 1637, recorded the cryptic vote, "Abraham Shaw is granted half of the benefit of coals or iron stone, which shall be found in any ground which is in the country's disposing."[6] Resident first at Watertown, and later at Dedham, Shaw figures in other surviving data only as a man whose house in the latter place had been burned (with some overtone of suspicion in Winthrop's *History* that he deserved this visitation from the Lord). Savage, mentioning the grant of mineral rights, says of it only that "this would more stimulate his curiosity than increase his wealth, had he not died the next year. . . ."[7] No evidence to indicate even the area in which he proposed to work has survived.

Thus, for all the careful advance planning, and however basic to economic self-sufficiency, the first efforts to develop New England's mineral resources must be pronounced failures. Perhaps it was a matter of first things first. The normal development of settlement, the work of clearing the land, building houses, and finding food may have absorbed all of the Puritans' energies. After all, mineral exploitation is uncertain and costly under the best of circumstances. It is

[5] James Savage, *A Genealogical Dictionary of the First Settlers of New England* (4 vols., Boston, 1860–62), II, 296; III, 144.

[6] *Mass. Records*, I, 206.

[7] Savage, *op. cit.*, IV, 63.

not at all unlikely, however, that the weak emphasis on metal development in the 1630's was the product of more than this, and of more than the exodus of the experts. Down to the outbreak of the civil wars, the regular coming of emigrants' ships served to keep the Massachusetts Bay colonists' needs well supplied. The need for selfsufficiency, on which the leaders had been counting in the early planning stage, did not, in fact, develop. The decade of the 30's was a boom period, with continuing increment of settlers, supplies, and new capital stimulating all kinds of business activity. It was only when these happy conditions ceased, owing to a change in the Puritans' lot at home in England, that the settlers here were driven to vigorous development of their own resources and the reactivation of old plans including work in metals.

It was, in other words, a depression which lay back of really vigorous efforts in lumbering, shipbuilding, and industry, homespun and otherwise, in the second decade of Massachusetts settlement. The impact of suspension of importation on the economy may be seen in an item in Winthrop's *Journal*, "The general fear of want of foreign commodities, now our money was gone, and that things were like to go well in England, set us on work to provide shipping of our own. . . ."[8] When things did go well in England, that is, when the Puritans there put through their political and military successes, immigration slowed down to a trickle, emigration set in, and New England was face to face, first with a depression, and second with the necessity to develop its own productive capacity for many things which, up to that time, had been steadily imported. As Victor S. Clarke has pointed out, while prices of land and provisions soared, thanks to increased demand from ever larger numbers of people, it was natural to exchange them for imported goods. When, however, the rate of growth declined, prices so sagged that there was nothing to exchange for such commodities as still came in. There was thus good reason to turn to the region's own resources.[9] While neither Winthrop nor Clarke specifically mentioned the start of metallurgical operations, in this context it seems clear that it fell in the general setting they described.

[8] Winthrop, *History of New England*, II, 29.
[9] Victor S. Clark, *History of Manufactures in the United States* (3 vols., New York, 1929), I, 130.

The depression-induced spurt to local self-sufficiency was, of course, by no means confined to metals. In area after area the new conditions stepped up production in fields started in the earliest period of colonization, and gave birth to new developments of raw materials and finished commodities. Ships of fair size had been built in the Bay Colony in the first years of settlement—the *Blessing of the Bay* at Mystic (now Medford) in 1631, the sixty-ton *Rebecca* at the same place in 1633, and a ship of one hundred and twenty tons at Marblehead in 1636—but from 1640 there was a major flurry of shipbuilding in Boston, Dorchester, Salem, and Cambridge. While glass is claimed to have been made at the village of Germantown, in Braintree, at a very early date, effective production began at Salem in 1639. Saltmaking had been coupled with ironworks in Graves' contract. A plant had been set up in 1636. The General Court, however, was pressing work in this field vigorously by monopoly grants to Samuel Winslow in 1641, and by concessions to John Winthrop the Younger in 1648, and to others still.[10]

In textiles, the development is clearest of all. Here the Court ordered, in 1641, a survey of "what seed is in every town, what men and women are skilful in the braking, spinning and weaving . . . ," etc., of linen and cotton.[11] Bounties were later offered and concrete results were forthcoming. Woolen manufacture got its start at Rowley about 1640, at the hands of Yorkshire immigrants, who also turned their attention to the making of cotton and linen cloth, with continuing support from the magistrates. Hubbard summarized the way necessity at first introduced what their "jurisprudence" afterward cultivated—"They soon found a way to supply themselves of (cotton) linen and woolen cloth."[12]

The rightness of general conditions has as much to do with the development of a successful industrial enterprise as the right people and the right resources. Here was a willingness, indeed an eagerness, on the part of the magistrates to try to fill the need by grants of monopoly and subsidies in various forms. What had been done to encourage other infant industries could readily be counted on for the by now ten-years'-neglected business of making iron from Massachu-

---

[10] *Mass. Records,* I, 331; II, 229.

[11] *Ibid.,* I, 294.

[12] Quoted in J. Leander Bishop, *A History of American Manufactures from 1608–1860* (2 vols., Philadelphia and London, 1864), I, 298–99.

setts ores. What was needed was the right man, and capital. The former was right at hand. In relatively short order he was to raise the funds, recruit the necessary workers, and get down to business.

By all counts, John Winthrop the Younger, son of the governor of the Massachusetts Bay Colony, was the right man to develop New England's iron resources, even though in retrospect his work in the exploitation of metals seems dwarfed by his accomplishments in a host of other fields. Universal genius, philosopher, scientist, and statesman, a "modern" among Puritans, a born leader, eventually the long-time and much-revered governor of Connecticut—the list could be extended almost indefinitely. Of strongly practical bent despite a life-long interest in esoterica, of marked business acumen despite a record of financial failures, energetic, able, persuasive, and popular—the project for a native Massachusetts iron industry could not have had a better "promoter."

Born at Groton Manor, the family estate, in 1606, educated at Trinity College, Dublin, trained in the law at London, and broadened by a European tour, he joined his father in Boston in the late fall of 1631. Chosen an "Assistant" almost immediately, he became one of the planters of Ipswich. On the death of his wife, he returned to England, but by 1635 he was back in Massachusetts as advance agent, on a one-year commission, for the Connecticut colonizing ventures of Lords Say and Sele and Brook. A beginning made in the establishment of a fort at Saybrook, he returned to Ipswich. There, with the new wife he had brought from England, a stepdaughter of the Reverend Hugh Peter, he took up again his magisterial responsibilities and prepared to lead the life of a respected country gentleman. It was not to be. Official duties at Boston and private concerns arising from an almost insatiable curiosity and a tremendous passion for projects guaranteed that he could not long remain quiet in this little settlement of which he was the acknowledged leader, or, for that matter, anywhere else. Now it was a matter of saltmaking, the setting up of the little works at Beverly. Unsuccessful, they caused a year and a half's absence from Ipswich. Now and again it was a questing for minerals, with many trips devoted to scrutiny of the new country's natural resources, to prospecting for ores of all kinds. And when at home, it was a common enough thing to be busy in the analysis of specimens sent to him by hopeful settlers scattered through Massachusetts and elsewhere.

Of his competence in theoretical and applied science there can be little question. Next to nothing is known of the content of his formal training in these areas, but it was at least the basis for a library whose range and calibre made it, in a lifetime of collecting, the best collection of scientific materials available in one place in America. Much of it has survived and may be readily inspected.[13] Of this nucleus half is made up of scientific works, the remainder of religious, historical, legal, linguistic, philosophical, and occult items. Half were written in Latin, the rest in English, French, German, Dutch, Greek, Spanish, and Italian. Here most of the great figures of science are represented—Paracelsus and Dorn, Dee and Glauber, Valentine and Stirk, and many more. Much is alchemical, but if an Agricola's *De Re Metallica* is not to be found in what has come down to us, there is a Cesalpinus *De Metallicis* and an Ubaldo *Le Mechaniche* to match a Mathesius *Bergpostilla* and a Buisson *Hercules Chymicus sive Aurum Philosophorum Potabile*.

Winthrop's scientific talents are also indicated by the range of his correspondents, scholars in many fields resident in many countries. He was to die with the reputation of having discovered the philosopher's stone. In life he enjoyed personal contacts and exchange of letters with an impressive group of men, including some whose greatness has been recognized by posterity for work vastly more practical than the quest for the means of transmuting base metals into gold. Among them are Comenius, Sir Kenelm Digby, Sir Christopher Wren, Prince Rupert, Sir Isaac Newton, and Robert Boyle. Eventually, too, he became a fellow of the Royal Society of London for Improving of Natural Knowledge. There he read papers on a wide variety of subjects, among them shipbuilding in New England, a self-feeding lamp, beer brewed from "Maiz-bread," and even "the tail of a rattlesnake which increased every year by one ring; whence the people conclude the age."[14]

These topics illustrate what was apparent to all who knew him. Winthrop was no "library scientist." He could ponder the strangest theories of the alchemists and eagerly buy each new work which seemed to promise answers. At the same time he could put his chemical, or alchemical, knowledge to work as a practising physician. All

13 The collection is owned by the New York Society Library.
14 Lawrence Shaw Mayo, *The Winthrop Family in America* (Boston, 1948), 51.

through life his diagnoses and prescriptions were eagerly sought, by letter and in person, by people of high and low status, scattered through the whole area of settlement, Indians as well as whites. The zest with which he handled these matters medical, even when busy with affairs of state and the concerns of private business, can be savored by even a casual reading of his letters in the *Winthrop Papers*. And when the residents of one town after another begged him to settle with them, often offering sweeping inducements, it can only have been because, in addition to his practical bent, he had the stuff of practical leadership and the ability to get things done.

Where Winthrop became acquainted with the manufacture of iron is not clear. He was doubtless familiar with the standard books on the subject, despite what a modern specialist in the history of iron technology would regard as lacunae in the surviving portion of his library. Nowhere is there anything to suggest that he had had practical experience in the field. It is not at all unlikely, of course, that his keen curiosity, and interest in mineral affairs generally, had drawn him at one time or another to watch ironmaking operations in England or in Ireland. Such observation, even by a man who saw much quickly, would hardly have provided more than a superficial knowledge of a tricky and difficult art. Winthrop was not, and doubtless never claimed to be, a real iron expert.

Our suspicions on this score are partially confirmed by one piece of documentary evidence. This is an item in the *Winthrop Papers* entitled "Sir Charles Coote's Account of His Ironworks," which describes several types of ore, including bog ore, lists the piece rates at which workmen were paid, and concludes with the caution, "Chiefly take care so to place your furnace that there be no water springs or damps under her for it will spoil all which if your ground will not admit, you must make a false bottom with several pipes to carry away the damps and water or springs."[15] Undated, but in Winthrop's handwriting, it was probably taken down when he was considering going into the iron business in Massachusetts. In it, except for its Irish focus, there is little or nothing with which one might confidently expect a man who really knew the art to be anything but perfectly familiar.

From all we can learn, therefore, Winthrop's projected venture

15 *Winthrop Papers,* IV, 363–65.

into iron rested on his book learning, some laboratory experiences, some presumably brief but doubtless keen observations of others at work. There was probably one thing more. As in other projects which he was to undertake during his lifetime, he may here have intended only to provide direction and leadership, to let imported specialists carry the various actual functions of erecting the works and making the iron. After all, ironmaking was no new art and it was hardly necessary for him to figure out his own way to build a bridge from New England ore to cast and wrought metal. A man of Winthrop's talents could provide the managerial skills for such an enterprise, even though his acquaintance with the ongoing work was only that of a highly gifted and, of course, much interested amateur.

There were other assets to offset whatever practical deficiencies this universal scientist bent on becoming an industrialist may have had. He had at least a fair estimate of New England's physical resources. He knew the area as well as any man there resident. He doubtless was fully aware of the extent of the Colony's need for iron as prospects of continuing importation grew dimmer. As his father's son and as a magistrate, participant in the councils of the policy makers, he well knew how far the General Court had been willing to go to encourage other infant industries. His own sponsorship of an ironmaking venture, given his position and his reputation, would certainly have guaranteed that the magistrates would go at least as far, and doubtless farther.

Much more was involved in the establishment of ironworks in New England than the successful surmounting of technological problems and the securing of assurance of concrete support from the local authorities. Ironmaking was a costly business. Even the basic physical plant required a major capital investment. Such an investment was demonstrably beyond the reach of either private purses or public treasury in Massachusetts, particularly at this juncture where needs were greatest and capacities weakest. The capital had to come from the Mother Country. To raise it would take no little quantity of salesmanship. No one was better qualified for this role than young Winthrop. Here, again, position and reputation would carry much weight, but so, also, would connections, if we may use today's term for a phenomenon as old as man. Winthrop could count on his family's and his own friendship with many members of a Puritan aristocracy of trade and commerce, on his relationship by blood and marriage to some of the rising stars in the Puritan political firmament. Add to these his brilliance, his eloquence, and his charm, and one would have to conclude that Winthrop was the perfect promoter for ironworks or anything else.

In June, 1641, the General Court passed its first measure having to do with work in metals since the vague grant to Abraham Shaw. The title in the margin of the order read: "Encouragement to discovery of mines, &c." The measure itself went as follows:

For encouragement of such as will adventure for the discovery of mines, it is ordered, that whosoever shall be at the charge for discovery of any mine within this jurisdiction shall enjoy the same, with a fit portion of land to the same, for 21 years to their proper use; and after that time expired, this Court shall have power to allot so much of the benefit thereof to public use as they shall think equal, and that such persons shall have liberty hereby to purchase the interest of any of the Indians in such lands where such mines shall be found, provided that they shall not enter any man's property without the owner's leave.[16]

That the hand of John Winthrop the Younger lay back of this act of the Court is nowhere made explicit. One must suspect that it was the result of proposals he had made, that it reflected both high hopes from such metallurgical data as he had been able to provide from his prospecting efforts and an awareness of the Colony's need for metals. It is interesting that iron is not mentioned, that the terms are somewhat vague, and that the order as a whole is a general one, valid for all who would see fit to risk money in Massachusetts' mineral prospects. One would give much to know the details of the negotiations which gave rise to it. We can only conjecture as to why it took the form just cited. If a particular site had been involved, it made obvious sense not to publicize it in advance of actual operations. It is conceivable that Winthrop, as well as the Court, had mineral ores other than iron in mind, perhaps the graphite in which he had been, and was long to be, interested. Finally, it was not yet clear who was actually to invest in the Bay Colony's still unexploited mineral wealth. One group or several, as many, indeed, as Winthrop's persuasion might influence, could feel adequately covered by the privileges here being offered.

16 *Mass. Records*, I, 327.

Precedents for such action were ample, and pressures, particularly in this period of capital shortage, strong. To encourage local industry the magistrates had offered bounties, as those on cloth made from native wool ordered in 1640;[17] land grants, as in the case of a number of gristmills starting with that at Watertown in 1635,[18] the glass house at Salem in 1638,[19] and the like; and even subsidies, as when the General Court asked Salem to lend £30 to the glassmakers, which would then be deductible from its next colony tax.[20] In all this, either particular local needs and particular local prospects or the general welfare of the colony, especially as threatened in the new crisis of general "depression," had been involved. Of this 1641 mineral-rights order the magistrates would doubtless have said that it was part of a general program of calculated "stimulation."

The encouragement in this case was hardly lavish. It went beyond that given Shaw in making an outright grant of land and full rights to minerals for a designated period rather than "half the benefit of coals or iron stone" which he had got. (One wonders, of course, whether it was the new economic situation or the involvment of a Winthrop which was responsible for the more generous inducement.) Even so, this grant, like Shaw's, seems to have meant to cover common lands and to protect private property owners. It also provided for an eventual umpirage to strike a balance between the interests of the private exploiters and those of the community at large. For all this, it would doubtless produce a very favorable impression on prospective adventurers of capital, particularly those in England to whom Winthrop proposed to turn.

Armed with this offer of support from the General Court, and doubtless with mineral specimens, John Winthrop set sail for England in the summer of 1641. The trip was roundabout and slow. It took fourteen days to reach Newfoundland. There he waited three weeks before a small ship bound for England could be found. The actual transatlantic crossing ran to twenty days of foul weather, and it was only on the 28th of September that the ship reached Bristol.

As far as Newfoundland, and possibly for the whole journey, young

[17] *Ibid.*, I, 316.
[18] William B. Weeden, *Economic and Social History of New England, 1620–1789* (2 vols., Boston and New York, 1891), I, 271.
[19] Essex Institute, *Historical Collections*, XVI, 2.
[20] *Mass. Records*, I, 344.

Winthrop was traveling in distinguished company. The magistrates had seen fit to send to England, for the purpose of looking out for Massachusetts interests in a changing political situation, Hugh Peter, pastor at Salem, Thomas Weld, pastor at Roxbury, and William Hibbins, of Boston. Peter, stepfather of Winthrop's wife and promoter of shipbuilding at Salem, became so caught up in the Puritan cause in the old country that he was eventually executed for treason when the Royalists triumphed. Hibbins, a substantial merchant, returned to Boston the following August, the immediate mission of the three emissaries accomplished. Weld, like Peter, never returned to New England, choosing instead to take a living at Durham, there serving God and the Puritan cause to the end of his life. He is known to have had a financial stake in Winthrop's ironworks. Perhaps the skilled promoter had put some of his time during the long rough voyage to good use by selling his proposed iron venture to his fellow passengers.

In all likelihood, these representatives of the Massachusetts Bay Colony, chosen as much for their connections with the Puritan party as for their ability, helped Winthrop make contact with some of the English capitalists whose financial support he had come to solicit. Early in 1642 he obtained still another useful ally in the person of his uncle, Emmanuel Downing. Then in London on private business, this man had built up, as a lawyer and merchant in the City prior to his emigration to Massachusetts, an impressive acquaintance among English merchants of Puritan sympathies. Since at this point the affairs of Massachusetts were at low ebb, with a calamitous financial depression plaguing the Bay settlers, and with Laud's *quo warranto* proceedings threatening the very existence of the Colony, it was all to the good of the proposed ironworks that Winthrop had such stalwart allies and intermediaries.

Details of the steps taken to persuade English investors that Massachusetts iron had the promise of becoming good business are few and far between. Given the care with which the Winthrops saved their letters, and the scarcity of references in the surviving collection to ironworks capital raising, one would conclude that essentially all of the negotiations had been handled on a man-to-man basis. It is easy to conjure up a picture of a persuasive speaker holding forth on the rich promise of New England's metallic resources, now, perhaps, in the smoke-filled taproom of a London inn, now in the crowded office of a merchant. For the quality of the ore, Winthrop could vouch as a recognized man of science. If any needed to be told of how quick and great had been the flowering of God's colony in Massachusetts, Winthrop could speak with vigor and authority. If any doubted the extent of the potential market for household ironwares and wrought iron, this man who had traveled far and wide in the whole area of settlement and knew what was needed and for what purpose could set them straight.

For all this, the job of persuasion took far longer than Winthrop had originally anticipated. His letters to his family back home carried frequent word of his regret that he had been so long away, of his hope for speedy return. Actually, it was more than a year and a half after his landing at Bristol before he set sail for home. Some of the interval, we know, was spent in travel on the Continent, but whether in quest of additional data on the technology of ironmaking or in pursuit of his general scientific interests is not clear. We know that in the fall of 1642 he visited Hamburg, Brussels, and The Hague, apparently visiting and corresponding with scholars. En route between Hamburg and Amsterdam, a chest of clothes, books, and other things was taken up by the men of a Dunkirk ship, opened, and its contents lost. Had an inventory been included in the scattered data on the lost chest that survive in the *Winthrop Papers*, we might today have a clearer picture of the purposes of this European trip and the role iron may have played in it. It is perhaps too fanciful, but a modern might well wonder if Winthrop had managed, in the course of this junket, to visit the Walloon country in which the ironmaking process he proposed to use in Massachusetts had first been developed.

Early in 1643, Winthrop was back in England, and very much engrossed in matters financial. In March he sold his property in Topsfield and Ipswich to a London "merchant taylor" for £250, doubtless to increase his own stock of cash. In the same month the embryonic industrialist, together with Peter and Downing, gave bond to secure Nicholas Bond's investment in the proposed ironworks. At different points and from various people, more probably during 1643 than in 1642, though actual data have not survived, came the capital subscription that finally took shape in the "Company of

Undertakers of the Iron Works in New England," the end product of Winthrop's capital-raising project.

Besides finding money, Winthrop had had also to recruit workers and take on a stock of materials and tools. Again, detailed data are lacking, but it looks as though the securing of skilled workmen had been every bit as difficult as the fund raising. Joshua Foote, a London ironmonger, had been able to scurry up three workmen, one a founder accompanied by his son. Three more, their names and points of origin unknown to us, ran away, perhaps from doubts as to the prospects of the new life stretching before them in Massachusetts, perhaps from boredom and lack of earnings as their departure was longer and longer delayed. Over these latter items those who stayed did much grumbling. Neither Foote's efforts nor those of Robert Child, friend of Winthrop, doctor of medicine, and universal scientist, had produced a bloomer, even in the four months following the departure of Winthrop and his band of workers. The Governor's son could hardly have taken much solace from Foote's suggestion that he "join all (his) workmen's heads together and see to build up bloomeries," even when it was coupled with the reassurance that "A smith after a little teaching will make a bloomer man." On those workers who actually sailed, we have essentially no data. We cannot tell how many there were or, except for the founder, what special trades they practised. None can be identified by name. Similarly, we may infer that there had been at least some trouble in assembling a stock of tools and materials, since arrangements for the purchase of stone, presumably for hearth and furnace lining, were not complete when the group sailed. Again, however, we do not know what items had been put together for the great venture.

By the 5th of May, affairs were well enough in hand to permit Winthrop to engage passage, on a part-payment basis, for the carriage of his men and goods to New England. On a day between the 20th and the 31st of May, 1643, Winthrop's party finally embarked on the ship *An Cleeve,* of London. His difficulties were by no means over, however. Held up at Gravesend by customs officials, he missed the favorable winds. Six weeks of waiting off the English coast were followed by fourteen weeks of tortured sailing across the Atlantic, during which Winthrop, his workmen, and his servants fell victim to ship's fever. When *An Cleeve* reached port in New England, it was late fall. The length of the voyage and an interval of convalescence from the illness meant that no real start with ironworks could be made until the following spring. Winthrop felt "damnified" by the actions of the customs people to the extent of at least £1000, and, in a document which carried the details just cited, petitioned Parliament for reimbursement. There is no record of action on the petition. It is clear from the paper in question that New England's ironworks had got off to a rather strained start.[21]

Despite all the difficulties, however, Winthrop was on these shores with at least some of the men and materials needed to begin the work of setting up furnace and forges. Back of him lay a group of men of substance whom he had persuaded to risk money in his undertaking and who could be counted on for additional financial and technological assistance. Ahead lay the bog iron ore deposits awaiting exploitation and conversion into the iron products of which a well-populated colony stood in need. The future was as bright as courage, ambition, and luck could make it.

[21] For details covering this preliminary period, see *Winthrop Papers*, IV, 341–425.

4

Puritans and Capitalists

THE COMPANY OF UNDERTAKERS of the Iron Works in New England was a private joint-stock company, in which eventually more than twenty people are known to have held shares. No copy of the original articles of agreement among the shareholders has been found. We know the identity of most of the partners. In some cases we know the size of individual holdings and something of how they were bought and how they changed hands. It is clear that many of those who risked their capital, at John Winthrop's urging or otherwise, had interests—business, religious and political—which touched at many points. These common interests made the business unit a far cry from the modern impersonal joint-stock company. Nevertheless, there is reason to believe that this Company was, in certain respects, an interesting transitional form between the simple private partnership and the modern industrial corporation.

It was not a corporation. Incorporation was definitely uncommon in the seventeenth century. It was a jealously guarded prerogative of states; it was invariably coupled with obligations of public service or benefit to the community at large; it was tightly defined in the rare cases in which it was granted. The Undertakers could not have secured the privilege of incorporation in England, since their enterprise was to serve, directly at least, the Massachusetts Bay Colony and not the realm of England. That it would provide "public service" to the

Colony was clearly recognized in its eventual receipt of monopoly privileges, land grants, and immunities from the General Court. The Company, however, was not incorporated in Massachusetts. There is no record that it ever sought the privilege. Had it done so, there is ample reason to believe that the magistrates would have declined to grant it.

In the first place, there was the problem of conflicting sovereignties. The Massachusetts Bay Company was itself a corporation, but neither its charter nor prevailing interpretations of common law gave it the right to set up private or public corporations within the limits of its patent. It is well known that Massachusetts took unto herself many of the rights of sovereignty. In the matter of incorporation, however, she seems to have trod very gently, probably from a carefully thought-out plan to avoid direct conflict with the parent government. Down to 1691, when Massachusetts became a royal province, the magistrates were to establish only two "private" corporations, a water company and Harvard College. They declined to set up Boston as a "public" corporation when its residents petitioned for the privilege. To incorporate a group made up principally of English residents would have been to invite trouble.

More important, the incorporation of straight industrial ventures was still far in the future. The first settlements, those of Virginia, Plymouth, and Massachusetts Bay, had, of course, been put through by, and at the expense of, public, that is, chartered corporations similar to the great joint-stock, foreign-trading companies. Their direct benefit to the realm was unchallengeable; their rights to incorporation could be readily justified. Under their aegis there were a number of joint-stock companies set up to carry through particular business and industrial operations. Among these were the Virginia glassworks, a project for "Transporting 100 Maids to Virginia to be made Wives," the "Magazine," and a "Society of Particular Adventures for Traffique with them of Virginia in a joint stock," all under the Virginia or London Company, and an abortive venture in which three trustees assumed trading rights and the control of certain property to be managed, under the Massachusetts Bay Company, for the benefit of investors who were to remain in England. No one of these colonial sub-ventures was chartered. They were, nevertheless, associations of businessmen which probably were closer analogies to the modern corporation than the state-chartered monopolies which were the parent Virginia and Massachusetts Bay companies.[1] Our data on these projects are so scanty that it is impossible to draw general inferences as to their organizational forms. If, however, we turn to certain German and Austrian business enterprises in the fifteenth century, we find forms which are closer prototypes of today's business corporations than anything in the Italian banking houses, in which customarily we seek the origins of capitalistic enterprise.[2] The practical pressures of a heavy-investment industry, of long-term and continuous operation, drove the capitalists of Augsburg, Nuremberg, and Leipzig, who controlled much of the mining activity of the Holy Roman Empire, to develop corporate forms in which share capital, continuity, delegation, and transferability were all present. These elements made possible the transition from the simple partnership, in which normally there was "full devotion of the participants' time, energy, and capital, equality in managerial power, freedom to terminate the undertaking on notice or for cause, and liability *in solido* for all debts incurred," to the "corporation," in which partial investment, part-time or "side-issue" interest, and limited liability were possible. The resultant business forms were what one might call *de facto* corporations, developed by businessmen to meet practical problems in the face of legal and political obstacles that obstructed formal incorporation.[3]

If the Company of Undertakers had something in common with the colonial sub-ventures mentioned earlier, it seems also to have had some of the characteristics of the *de facto* corporation. Investment was by transferable shares. Downing, for example, sold in October, 1645, for £60 sterling what he described as "all my right and interest as an vndertaker in the Iron Works, wherein my part and share is fifty pounds. . . ."[4] There was continuity. In the 1699 will of Edmund

---

[1] Joseph Stancliffe Davis, "Corporations in the American Colonies," *Essays in the Earlier History of American Corporations* (2 vols., Cambridge, 1917), I, 20–21, 39; Shaw Livermore, *Early American Land Companies* (New York, 1939), chs. 1–3.

[2] Jacob Strieder, "Origins of Early European Capitalism," *Journal of Economic and Business History*, Vol. II (1930), 1–19.

[3] I am here following Livermore's well substantiated general thesis that the *voluntary* business associations to which men turned in the face of legal and political obstacles to incorporation were the real "ancestors" of today's corporations. I am also giving the Company of Undertakers more importance, in these terms, than Livermore could with the data at his disposal.

[4] *Winthrop Papers*, V, 47.

Spinckes, drawn long after the ironworks had ceased to exist, a £50-share acquired in 1646 was bequeathed to his eldest son.[5] Actually, individuals were buying into, or selling out of, the Company during its whole span of operations, and the Spinckes bequest is by no means uniquely *ex post facto* among recorded transfers. The element of "side-issue" investment may be seen again and again in the short biographical sketches of the various investors which follow. Share holdings were unequal, ranging from the basic £50 of Downing and Spinckes to those of Gualter Frost and John Becx, which allegedly ran to £1500 and £2000, respectively.[6] In no case, however, was investment in the ironworks a main concern of the individuals making up the Company.

Whether or not formal arrangements for delegation of control were set up in the Company's articles is not known. It was certainly practiced. In normal times, there seems to have been nothing like a standing board of directors. Letters sent in the Company's name were never signed by the full complement of shareholders. Certain names almost invariably appear but there is much variety among the others. One of the Company's agents claimed that he had seen an "order" which "impowered five or three to be sufficient to act for the company in the management of their affairs," but the instrument has not been found.[7] When difficulties were encountered, however, there was a delegation of control which can be documented. In 1652 eleven "Undertakers and Copartners amongst others" set up three Massachusetts residents, presumably all shareholders, as "True and lawfull deputies and attorneys for (the eleven) & all other the said Vndertakers & aduancers." Five years later, eight "Adventurers," only five of whom actually signed the document, gave John Becx sweeping power of attorney to act for the Company.[8] Since the broad Atlantic separated the capitalists from their investment, actual management had to be by delegation. In the case of two agents, Winthrop and his successor, Richard Leader, the men who ran the Company's American operations were in all likelihood both share-

---

[5] Henry F. Waters, *Genealogical Gleanings in England* (3 vols., Boston, 1885–89), 171–72.

[6] Ironworks Papers, 228–29.

[7] *Essex R. & F.*, III, 43.

[8] *Suffolk Deeds*, I, 229–30; III, 157–58.

holders and paid managers.[9] The third agent, John Gifford, had no share in the venture and therefore much resembles the hired managers of modern corporations.

The Undertakers sued and were sued as a company, although normally, and doubtless because of the lack of formal incorporation, the names of Becx and a number of others were coupled with the general designation of "Company of Undertakers." It held and could dispose of its own property, apparently without the specific consent of all shareholders. There is no evidence of disagreement among shareholders as to the carrying on of the business by an "executive committee," so to speak, until after the Company got into financial difficulties. Finally, and perhaps most important of all, the Company seems on one occasion to have met and passed the crucial test of limited liability.

In the tangle of litigation arising from the confiscation of the Company's estate for debt in 1653, a local court verdict went against the Undertakers, and execution was ordered on the private property of Henry Webb, a shareholder in, and attorney for, the Company. Webb challenged the validity of the execution. Eventually, the General Court found that he could not be held personally responsible. It held that Webb was indeed a shareholder, but pointed out that he had not signed articles with the Company and was merely an "adventurer for profits and loss only and refusing to be responsible . . . further than the money goes." The magistrates seem thus to have recognized a distinction between simple investors and full-fledged "Undertakers," which may have been good law or which may have been tailor-made to the particular situation. The decision was not easily reached. Recorded in opposition were some of the best legal and business minds in the colony. Neither does it seem in any way to have served as a precedent in the subsequent growth of Massachusetts jurisprudence. Though lost to history, the Webb case is an American "first" of major significance. If, as it appears, the magistrates were here painfully groping their way toward a distinction between investors and what we would call the promoters of a business, they were far in advance of their time.[10]

While there are ample precedents of simple partnership in the

[9] *Winthrop Papers*, V, 5–8, 14, 21–22; Suffolk R. & F., 125–2.
[10] *Mass. Records*, III, 351; IV–1, 188; Mass. Arch., 38–B/33.

English iron industry, and at least one of a more complex partnership in which the germs of corporate practice were present, the Company of Undertakers seems to have been an unusually sophisticated business organization. It has been claimed that coal and iron mining had been "the earliest stimuli to the development of true joint enterprise along the lines of the modern corporation" on the Continent and in England.[11] Iron, that most basic of metals, was the foundation for what may well have been the first successful and large-scale corporate enterprise, apart from the colonizing companies proper, in America. It is doubtful, of course, that the business form and business procedures had been especially contrived for the job in hand. They doubtless just grew, but their growth was eminently logical. Given the scale of operations envisaged and the separation of the shareholders from their physical investment, the simple partnership, however extended, would have been wholly inadequate.

However many of the tests of the corporation the Company of Undertakers may have passed, it fell far short on that of impersonality. The investors probably all knew one another. Certainly, they had many things in common. Of the twenty-four known shareholders, ten were London merchants, three were administrators, two were merchants from other parts of the country, two (one of them already an emigrant to New England) were lawyers, four were Puritan ministers, and one, Robert Child, so versatile as to be almost unclassifiable. Four of them, Copley, Foley, Becx, and Foote, had firsthand knowledge of the English iron industry; two others, Pury and Beeke, may have known something about it. To the rest, so far as we can tell, a share in the Company of Undertakers was just another investment. Where ordinary business acumen was called for, their advice may have been valuable, but they had no technical knowledge. In wealth, the group ranged from the very wealthy to the moderately substantial. Three served, sooner or later, as members of Parliament. Many had fairly distinguished records of public service.

If this gives an appearance of variety, there is no doubting a common bond of religious sympathy. Only eight, obviously including the four clergymen, can be positively identified as Puritans. Since the initiative for building the ironworks came from the colonists,

[11] *Sussex Arch. Coll.*, XXXII (1882), 31–32; C3/439/39; C5/22/27; C8/185/69; Livermore, *op. cit.*, 41.

with Winthrop as the link between them and the investors, it was inevitable that he turn first to people whom he, his family, and friends knew well. These would have been, without doubt, not merely merchants but Puritan merchants. The circle may have been broadened a little. Two of the shareholders were members of the Dutch Reformed church, for example. There is little doubt, however, that, of those whose religious persuasion cannot be ascertained, all had more or less strong Puritan sympathies.

Interestingly enough, the principal figure in the Company was one of the Dutch Reformed church members. In many surviving documents the firm is designated as "John Becx and Company of Undertakers of the Iron Works in New England." Becx, a Dutch resident alien of London, and one of the four shareholders with ironworks experience, was a wealthy man and a typical seventeenth-century merchant industrialist.[12] His business interests were far flung and varied. Apart from his large holdings in the Company of Undertakers, he owned land and a large sawmill in New England and traded with both New England and the West Indies. At one point he undertook to relieve Cromwell's party of the care of Scots taken prisoner at Dunbar and Worcester by buying their services at so much a head and selling them at a profit in the New World. He also owned ironworks—in the 1650's, in the Forest of Dean and Gloucestershire, and in 1665, in Ireland. The latter, located about thirty miles from Dublin, were not too profitable, because of the competition of Spanish and Swedish iron. Once, too, he asked for the chief position in the post office at London on the strength of the languages at his command and his skill at accounts; alternatively, he would have liked the equivalent post at Dublin. That some of his affairs prospered is indicated by the high rent he paid while living in Chelsea, and by the fact that he once loaned over £15,000, no mean sum for the middle of the seventeenth century, to Sir Thomas Ashfield, a Royalist who was heavily fined by the Commonwealth.[13]

His letters summon up a picture of a man weighed down with business concerns, with whose details he wrestled with enormous

---

[12] W. J. C. Moens, *The Marriage, Baptismal and Burial Records . . . of the Dutch Reformed Church, Austin Friars* (Lymington, 1884), 4, 76, 212; *Journals of the House of Commons*, VII, 433, 453, 460.

[13] C5/591/140–41; C24/1080.

energy. The vigor of his prose testifies to more than a mere "command" of English, and the carefully worked out "forms of account" which he sent to the ironworks factor and agent as models for their bookkeeping suggest that he had no little skill at accounting.[14] Completely dominated by the profit motive, he was quick to seize an investment opportunity, not troubled by too-strong scruples, and, first, last, and always, brash. Even after disaster had struck the Massachusetts ironworks, and his fellow shareholders had declined to continue to pour good money after bad, he was capable of trying to work out an arrangement for joint operation with the creditors to whom the Company's assets had been handed by the courts. He was capable of writing as he did so, "Had I the managing of those works in my hand I would as easy as I turn my hands make £2000 per annum profit . . ."[15]—a figure which, in the light of experience of the works to that date, was literally fantastic.

In the very beginning of the venture, Becx's role was secondary to that of Lionel Copley, to judge from the fact that the latter's signature came first or second on important letters dispatched to America by the Undertakers. One of the principal ironmasters of the period, Copley's interests, too, were broadly extended. The basis of his family fortunes was his position as a country squire. In one of his lawsuits he described himself as "a great dealer for many thousand pounds per annum in wood, iron work, coal mines and other things. . . ." His time was considered so valuable that when employed to negotiate a lease for two other capitalists that involved travel from Yorkshire, he received a commission of £100, a large sum, for "neglecting his own affairs."[16] His coal mining interests probably centered, like his ironworks, in South Yorkshire, Nottingham, and Derby. Dying in 1675, he left a will that made no mention of either ironworks or coal mines. It is clear, however, that he died a rich man.

Both his early prominence and later eclipse in the affairs of the Company of Undertakers can be explained by the vicissitudes of his political career. During the Long Parliament, he sat for the borough of Bossiny in Cornwall, and rapidly rose to high position in the Parliamentary Army. In July, 1643, he was a commissary; the following

14 Suffolk R. & F., 225.
15 Ironworks Papers, 233.
16 C10/92/28; C6/38/42.

year he was appointed colonel of a regiment of horse; by 1648 he was Commissary General of the army. In December, 1648, however, he was arrested, along with four others, on a charge of having participated in an invitation to the Scots to invade England. He remained in prison, untried, and uncondemned, for several years, obviously in no position to carry any part in the Undertakers' affairs. He was back in the picture after the Restoration, however, and, now a staunch supporter of the King against the Independents, took a leading part in the investors' last-ditch fight to recover their confiscated New England ironworks. His name heads the list of Undertakers who petitioned the King in 1661, and he appears to have been the guiding spirit in the abortive lawsuit which was brought before the New England courts in 1663.

Thomas Foley was another shareholder of no mean importance and with a background which well qualified him for an interest in Winthrop's venture. Of him Richard Baxter asserted that "from almost nothing, (he) did get about £5000 p. ann. by ironworks."[17] Born in 1617, the second son of Richard Foley, an important Worcester ironmaster who in 1616 was mayor of Dudley, he was to own, by the time of his own death in 1677, forges and a slitting mill at Hartlebury, in Gloucester, ironworks and mine works in Monmouth, shares in business partnerships in Gloucester, Hereford, and Monmouth, and large estates in Worcester and Hereford. One measure of his importance in the English iron industry is that, with George Browne, he was the executor of the will of John Browne, the armaments manufacturer. The two executors in all likelihood carried on the business after his death. Another is that, when, in 1665, Andrew Yarranton was sent on a mission to Saxony to discover the secret of the manufacture of tin plate, Thomas and Philip Foley were among his sponsors.[18] In religion, Foley was a moderate Puritan and a member of Richard Baxter's congregation at Kidderminster. His attitude toward the monarchy is uncertain.

Last of the group of men with experience in the iron trade is Joshua Foote, a London ironmonger. He was a prominent member of the Company of Ironmongers. Once a member of a six-man group ap-

17 Quoted in *D.N.B.* article on Foley.
18 Samuel Smiles, *Industrial Biography: Iron Workers and Tool Makers* (London, 1876), 66 n.

pointed to break a patent held by John Brown for making and selling all sorts of cast ware, considered "prejudicial to this Company and the commonwealth," he was, on two occasions, next to the highest man when it came to meeting forced loans to the government.[19] While trade in iron was his main concern, he was not unacquainted with its manufacture. He had been one of a group who set up an ironworks at Tomgraney, in County Clare, which was entirely merchant financed, and which may have been in several respects an Irish prototype of the Massachusetts venture.[20] He seems also to have intended to erect some kind of ironmaking plant of his own in New England. He emigrated to Massachusetts just in time to see the Company's ironworks in dire straits. With its residents he had had business dealings from as early as 1635, and, from 1644 until his coming, these had been quite extensive, including the lending of money on mortgages and the purchase of real estate in Braintree, Boston, Roxbury, and elsewhere.[21] Moving to Providence in short order, perhaps because by no stretch of anybody's imagination could *he* qualify for limited liability for the Company's debts, he died there in 1655, leaving an estate whose settlement involved long-drawn-out legal proceedings.

Of some of the shareholders in a category of general merchants we know next to nothing. On others we have quite extensive data. In the former group stands William Beauchamp, one of the original investors, and described as "of London, merchant."[22] Rowland Searchfield, another Undertaker, and apparently a Bristol and London merchant, may have begun life as a relative of a bishop of Bristol, of the same name, and as the apprentice of a Bristol alderman. Back in England by 1644, after a stretch as a merchant at Tercera in the Azores, he was, by 1655, living in London and importing oil from Italy. George Sharpulls, another London merchant, has been hard to identify. In one lawsuit in which the name appears, he was objecting to the trustee chosen to look after some legacies due to be paid to his children. The money might be lost, he alleged, because among other

---

[19] John Nicholl, *Some Account of the Worshipful Company of Ironmongers* (2nd ed., London, 1866), 231, 240, 257.

[20] C24/733/33; Rawlinson MSS. (Bodleian Library), D 918/133.

[21] *Winthrop Papers*, III, 208; *Suffolk Deeds*, I, 66, 153, 166, 167, 213, 328, 335.

[22] *Suffolk Deeds*, I, 229.

things the trustee had "dealings in parts beyond the sea" and might emigrate. Sharpulls implied that neither of these strictures applied to him, and that he would therefore make a much better trustee.[23]

John Beck, who signed this name "Jo. Beeck," was, like Becx, a member of the Dutch Reformed church. He was probably the son of a merchant of German birth many years resident at London, Abraham Beck. The father is characterized in one lawsuit as "in his lifetime a greate dealer and Trader in way of Merchantdize." Of the son we know but little. While serving as executor of his father's estate he paid or gave security on £14,000 of the latter's debts before discovering that, thanks to heavy losses in overseas trading, assets were far outweighed by liabilities.[24] This clearly was a case of a son's being embarrassed by his father's fine commercial reputation. The figure cited suggests that he must have been a man of some substance. In 1649, Beck was on the Committee of Militia for the Tower Hamlets, now the East End of London, but then a trading and wealthy residential area, and presumably a supporter of Parliament. The following year he was on the other side, or at least was being charged with concealing a delinquent's estate. From one bit of data produced in the proceedings, it appears that by that point Beck had become a merchant, trading to the Low Countries.[25]

Quite as though there were not already enough chance of confusion in the names Becx and Beck, there was also a William Beeke in the Company. This man, holder of a half share which ultimately came, with advances, to £350,[26] was a "Merchant Taylor," and quite a prominent one. Admitted to the Freedom of the Company in 1607, and a member of the Livery in 1639, he became a member of the Court in 1653 and Master in 1657. A London linen draper, wealthy enough to own, jointly with a partner, a house which cost £1200 in 1644, he was also a devout Puritan, and left £20 in his will "to be disposed amongst silenced ministers."[27]

Another "merchant taylor" who invested in Winthrop's project,

---

[23] C9/232/164.

[24] E112/218/1035.

[25] *Calendar of the Proceedings of the Committee for Compounding, &c., 1643–1660* (5 vols., London, 1889–92), Part I, 265.

[26] Suffolk Deeds, XVI, 42.

[27] Hustings Roll 320, no. 9 (City of London Record Office); Prerogative Court of Canterbury—Wills, Hene, 29.

John Pocock, had, through his rare combination of religious zeal and business capacity, an almost unmatched record of usefulness to the New England Puritans. An English member of the Court of the Massachusetts Bay Company, its agent in many pressing concerns in the old country, and a patient creditor, he served them as he was later to serve English Puritans during the Civil War and the Commonwealth. His stature is indicated by his rise in the Company of Merchant Taylors, to which he had been admitted in 1615. He did not obtain the dignity of Master, but he became a member of the Livery in 1624 and of the Court in 1641, fourth Warden in 1643 and second, in 1644. In 1633 he described himself as a woolen draper. In the 1640's he supplied the New England colonists with cloth worth £150 and had great difficulty in obtaining payment, the last of four installments being paid only in 1655.[28] In 1643 he supplied Parliament with coats, presumably for the army, worth £428.10s.; on another occasion we find him supplying cloth to other London merchants.[29] This makes it difficult to tell whether he was normally a wholesale or retail merchant, and whether he dealt in cloth or clothes. In any event he was a quite prosperous man.

An ardent Puritan, who had in 1627 contributed to a fund to buy up advowsons and replace orthodox ministers by Puritans, and had helped to found the Congregational church at Stepney, he became treasurer of a fund for maimed soldiers and war widows and orphans, and a trustee of the fund for Ministers' Maintenance. These services in behalf of the Puritan cause at home were the complement of his work as "commissioner" of the Massachusetts Bay Company in England, for which he received at least a vote of thanks.[30] With little question, one of the first prospective investors to whom Winthrop would have turned, he is almost the perfect type for carrying a share in a venture which would advance the welfare of a colony to which he never went but whose cause he ably supported through the years.

Of those investors whose general business interests centered in textiles, there is, finally, one of whom we know little indeed. Samuel

[28] C8/53/169; *Mass. Records,* II, 82, 138, 262; III, 48, 144, 247–48, 255, 291; IV–1, 74; and R. P. Stearns, "Letters and Documents by or Relating to Hugh Peter," Essex Institute, *Historical Collections,* LXII (1936), 55–57.

[29] Historical Manuscripts Commission, *Fifth Report* (London, 1876), 98a; C10/165/91.

[30] *Mass. Records,* II, 138.

Moody was presumably a wool draper of Bury, whose brother John emigrated to New England in 1633, and who left lands in Norfolk, Suffolk, and Ireland on his death in 1657. He was a "Justice of the Peace since the death of King Charles, and chosen by the borough of Bury into severall Parliaments in that time."[31]

The two brewers who were members of the Company are almost as obscure. William Hiccocks was a Southwark resident who was a Warden of the Brewers' Company in 1652, and its Master in 1657. A member of the Court of Assistants from the fifties until his death in 1674, he was not eminent enough to be an alderman of the City of London, being twice rejected, in 1667 and 1668. He left some land in Sussex and much leasehold and freehold property in Southwark. Robert Houghton, also a Southwark brewer, made beer which was good enough to supply the King. In 1637 he had a share in the ship *Christopher and John,* but his appearances in lawsuits are so few and uninformative that we are at a loss to know the character and extent of his estate and investing habits.

Data on the three administrators are far more extensive. One, Nicholas Bond, was a man of some prominence, sufficient to give his signature a place of priority in many of the Undertakers' letters, and to have him make it plain on one occasion that he was "Nicholas Bond Esq." and not "Nicholas Bond merchant." [32] From 1642 to 1646 he was controller of the household of Henry, Duke of Gloucester, and Princess Elizabeth. Then, the Parliament finding cause to reduce the household of the King's children, he was pensioned off at £200 yearly.[33] From 1646 he presumably remained at Whitehall as a civil servant. After 1650 he was regularly chosen as "steward" or organizer of state banquets to celebrate a day of thanksgiving or to honor visiting ambassadors. In the same period he was also a member of the committee for taking the Commonwealth's accounts, at a salary of £200, paid in Irish lands rather than in cash.[34] Aside from his share in the Company of Undertakers, which came originally to

31 H. F. Waters, *op. cit.,* 96–97.

32 C9/374/19.

33 *Calendar of State Papers, Domestic Series, of the Reign of Charles I* (23 vols., London, 1858–97), 1645–47, p. 247.

34 *Calendar of State Papers, Domestic Series, 1649–1660* (13 vols., London, 1875–86), 1652–53, p. 157.

£100,[35] he invested in land, took part in the lotteries for Irish land, farmed the excise of timber for some years prior to 1658, and lent small sums to merchants and others.

His fellow shareholder, Gualter Frost, appears in a contemporary account as "Mr. Walter Frost, the Manciple of Emmanuel College, since swordbearer to the Lord Mayor and afterwards Secretary to the Councill of State, a man beyond all exception for integrity of life, an excellent mathematician. . . ."[36] He was probably the "W. Frost" who published at Cambridge "A New Almanacke of Prognostication for the Yeare of our Lord God 1627." In 1645 he was appointed Chronologer of the City of London, called upon to collect and record "all memorable acts of this City, and occurrences thereof," at a salary of one hundred nobles a year. His immediate predecessors in the office were Ben Johnson and Francis Quarles.[37] At the outbreak of the Civil War, Frost took the side of Parliament and rose rapidly. Parliamentary Commissary in Ireland in 1643, the next year secretary to the Parliamentary Committee of Both Kingdoms, from 1649 until his death in 1652 he was secretary of the Council of State, at the good salary of £4 per day.[38]

Dying suddenly, he left no will. From the lawsuits which resulted, however, we can learn much of his business activities. In the claim of a daughter, he had died worth £12,000, the estate including personal property, shares in the Excise, the East India Company, the New England Company (the ironworks), the Gunpowder Mills, and Island Eleutheria worth £2000, £2200, £1500, £1400, and £300 respectively, and gifts from Parliament amounting to £1200. According to his eldest son, however, his father's estate had greatly depreciated, as when "the £1500 mentioned to be in the New England Company was every penny of it lost," and when similar tragedy had befallen the investment in Island Eleutheria. By his reckoning, the estate amounted only to £6520, but even this was a tidy fortune for the

[35] Suffolk Deeds, XVI, 42.

[36] Atwell, *The Faithful Surveyor* (1662), 81, quoted in Historical Manuscripts Commission, *Fourth Report* (London, 1874), 418.

[37] *Analytical Index to the Series of Records Known as the Remembrancia, Preserved among the Archives of the City of London, A. D. 1597–1664* (London, 1878), 305–306.

[38] *C. S. P. D. Charles I,* 1641–43, p. 475; Historical Manuscripts Commission, *Report on the Manuscripts of F. W. Leybourne-Popham, Esq., of Littlecote, Co. Wilts* (Norwich, 1899), 37; *C. S. P. D. Charles I, passim.*

middle of the seventeenth century.[39] Incidentally, the son eventually sold the interest in the Company of Undertakers, this time listed as a full share amounting, with advances, to £700, to one James Dewy, despite his earlier admission that the Company had failed.[40]

Cornelius Holland, like Frost and Bond, was a professional administrator. In 1631 a member of the Household of the Prince of Wales, by 1635 he was its clerk-controller and, as such, Bond's superior. It was presumably this bureaucratic contact which led him to invest in the Undertakers. The size of his interest is not known. He never loomed large in the Company's affairs. Clearly, however, he was enormously weathy. Noble, in his *English Regicides,* points out that "few, from so small a beginning, obtained such considerable grants as Mr. Holland," and goes on to show that "considerable" referred to lands worth from £1800 to £14,000 or £15,000 a year.[41] He also held a series of offices that can only have been lucrative. In 1642, along with Pocock and Greenhill, his fellow Undertakers, he held funds for the relief of maimed soldiers and of war widows. In 1643 he was appointed collector of the profits arising to the King from the mint. From 1650 to 1652 he was a member of the Council of State, and served thereafter on a commission appointed by Act of Parliament to "inspect the Treasury." He also had experience in colonial affairs, being named in 1643 one of the Commissioners for Plantations, and in 1653, one of a Company for the Somers Islands. According to Noble, he was worth at least £50,000.[42] Given his money and his political influence, he must have been regarded by Winthrop as presently and potentially a most useful member of the Company of Undertakers.

Thanks to the large role played by religion in the period with which we are dealing, we have a great deal of data on the gentlemen of the cloth who saw fit to invest in New England iron. Less ample but rather depressing data on their financial activities suggest that they would have done better to stick to their professional concerns. William Greenhill D.D., 1591–1671, was one of the leading lights of the Congregational cause. Educated at Magdalen College, Oxford,

[39] C7/461/40.
[40] Suffolk Deeds, XVI, 42.
[41] Mark Noble, *The Lives of the English Regicides* (2 vols., London, 1798), I, 358.
[42] *Ibid.,* I, 360.

and rector of New Shoreham, Sussex, from 1615 to 1633, he was one of the founders and first pastors of the Congregational church at Stepney in 1644, one of the chaplains to the King's children, and one of the treasurers of the fund for relief of maimed soldiers and of war widows and orphans. Removed from the living at Stepney at the Restoration, he stayed on as pastor of the Congregational church. Apart from his share in the Undertakers, apparently a half share coming to £350, including advances, we have few traces of his investments.[43] In 1642 he put £25 into a lottery for Irish land and secured property in County Down. In the same year he bought an eighth share in a two-hundred-ton ship for some £112, but twenty years later he had received no dividends. Suing to recover, he admitted his lack of experience in "maritime affayres." The defendant brazenly took refuge in the statute of limitations—a sorry comment on Greenhill's business acumen.[44] His will disposed of land holdings in four Norfolk and Suffolk villages. No investments are mentioned. Experience may have taught that such things were not for him.

Edmund Spinckes is less well known and figured little in the affairs of the Company of Undertakers. Graduated from Emmanuel College, Cambridge, he was rector of Castor, in Northamptonshire, from 1646 until 1661, when he became rector of Orton Longneville, in Huntingdonshire. He served there until deprived of his post in 1662 under the severer tests for Puritan ministers introduced in that year. No better tribute to his shortcomings in the world of affairs is needed than the item in his will mentioned earlier. If he believed in 1669 that his interest in the Company of Undertakers was still worth bequeathing, he had hardly followed the Company's fortunes with any careful attention.

Thomas Weld, the third clergyman investor, was born the son of a Suffolk mercer and educated at Cambridge. Vicar of Terling, in Essex, from 1624 to 1631, he was deprived by Laud for nonconformity, and chose to emigrate to New England. In 1641 he returned to England as agent for Massachusetts and found two parishes at opposite ends of the kingdom that needed his attention. We hear of him serving at St. Giles Cripplegate, London, in 1644 and 1647. In

---

[43] Suffolk Deeds, XVI, 42.

[44] *Calendar of the State Papers Relating to Ireland Preserved in the Public Record Office, 1625–(1670)* (8 vols., London, 1900–10), Adventurers for Land 1642–59, pp. 43, 343; C9/49/77.

the grants to the Company of Undertakers, he is described as "minister of Gateshead," where in 1649 he became rector. He and his parishioners did not agree, and in 1657 he published "A Vindication of Mr. Weld," in which he replied to their complaints in a very arrogant manner. Their chief grievances were that he refused to baptize infants and to administer the sacrament (as he put it) "promiscuously" to "liars and scorners," but insisted on reserving it for a favored few—ten in all![45] He died in 1661, but had, a good ten years before, presumably transferred his interest in the ironworks to relatives who had stayed on in Massachusetts.

The Reverend Richard Babington, the last of our clerical group, was the son of a London merchant. He studied at Christ's College, Cambridge, taking his bachelor's degree in 1625, his master's in 1628. Becoming vicar of Sidbury, in Devon, in 1630, he served there until ill health forced his resignation in 1650. He was one of the signers of the Testimony of 1648, a Presbyterian manifesto. In 1655 a man of this name who had farmed the excise on gold, silver, and copper wire at £2800 a year—a large advance—was complaining to the Commissioners for Appeals and Regulating the Excise that traders were resorting to fraudulent practices to elude duty. If this was the clergyman, it is clear that his extracurricular interests had been broader, and far more lucrative, than investment in the Company of Undertakers. He died in 1681 or 1682. In his will, proved in 1683, he left £100 for ejected ministers and directed that three Conformists and three Nonconformists should carry his coffin. Of him it was said that he studied "Physick" but gave advice free, that he had a good estate, and was a "learned and moderate man."[46]

Thomas Pury, M.P., one of two investors whose main concern was with the law, was described in a Royalist pamphlet as "first a weaver in Gloucestershire and then an ignorant country solicitor."[47] The terminology is unkind but Pury had made the somewhat surprising transition from a clothier to a lawyer. When in 1622 the clothiers of Gloucestershire and Oxford presented a petition to the Privy Council on the depression of their industry, it was Pury's name which headed

[45] *D.N.B.* article.

[46] Edmund Calamy, *A Continuation of the Account of the Ministers . . . Ejected and Silenced after the Restoration . . .* (London, 1727), 376.

[47] *The Mystery of the Good Old Cause.*

the list of signatures. By 1627 he was in practice as a lawyer. Sent to the Long Parliament from Gloucester City, he took a leading part in the defense of Gloucester against the King. After the execution of Charles I, he continued to support the government, and sat in Cromwell's first Parliament (1654–1655). Little is known of his private business concerns, but it is possible that knowledge of the iron trade may have gone with his skills in the clothier's and lawyer's arts. His son, at least, had ironworks, in partnership with others, near the Forest of Dean.

Emmanuel Downing, the other lawyer, has already been mentioned as one of the more useful of Winthrop's contacts with English merchant capitalists. Probably the son of an Irish clergyman, Downing was a lawyer of the Inner Temple, the brother-in-law of Governor Winthrop, and a most helpful intermediary between the Massachusetts colonists and the English government in the period between the Winthrop migration and his own removal to New England in 1638. Several of his letters on American topics have been printed in the correspondence of Sir John Coke, secretary of state.[48] In December, 1633, he wrote a stout defense of the Massachusetts Puritans, in which he pointed out that they had emigrated for conscience's sake and to convert the heathen, and were not contemplating secession. He had a keen nose for news and was an extraordinarily good reporter, talents which were doubtless as valuable to the young Colony as they were useful to Winthrop the Younger.

Welcomed here as an important and long-awaited settler, serving as a local magistrate and deputy from Salem, and awarded a six-hundred-acre land grant, he was at least twice called back to England on business. Long charged with the care of the Winthrops' private business matters in England, he could fairly have been expected to take a share in his nephew's ironworks venture. He did so, on credit, however, and after Winthrop's return to Boston. Staying on in London, he was able to report by letter on developments among the other shareholders, developments which presumably influenced him, at least in part, to sell his share, at a £10-profit, in October, 1645, as we have seen. It is just possible, however, that he continued to carry some stake in the ironworks.

[48] Historical Manuscripts Commission, *The Manuscripts of the Earl Cowper, K. G., Preserved at Melbourne Hall, Derbyshire* (3 vols., London, 1888–89), I, II.

Thomas Vincent, the purchaser of Downing's £50-share, was probably the "Mr. Vincent" to whom young Winthrop paid £50 in London for linen cloth which he had sent to New England.[49] A leatherseller, and in 1654 Master of the Leathersellers' Company, he also traded in wool as well as in linen. The combination of activities may seem strange, but it has at least one contemporary precedent.[50] The records seem otherwise to be silent on this man who never became prominent in the Company of Undertakers.

Last, but far from least interesting, of the shareholders is the gentleman who defies classification, the famous Doctor Robert Child. Born to a substantial Kentish family, he took his A.B. and A.M. degrees at Cambridge, studied medicine at Leyden, and received his M.D. at the University of Padua in 1638. Wide travels in Europe, which provided food for his interests in scientific agriculture, chemistry, metallurgy, and medicine, were matched by a survey of the New England settlements in the late 1630's. Convinced of the mineral and other economic possibilities of the Puritans' new land, he was ready indeed to invest in the great ironworks project when Winthrop again met up with him in England. He also put money into Winthrop's graphite operations at Tantiusques and into an ambitious fur-trading scheme. Returning to Massachusetts in 1645, he planned to set out vineyards, apparently in conjunction with ironworks in the Nashua River country, and also acquired huge tracts of land in Maine.

These activities all petered out when Child became the leading spirit in a movement to extend suffrage and broaden religious toleration in the Colony. Passions reached great heights in a controversy in which it was held that the very foundations of civil authority were threatened. Before Child permanently quit these shores in 1647, he had been twice brought to jail, once for a petition judged contemptuous of the magistrates, and once for sedition, had there spent three months of confinement, and had been assessed fines of £50 and £200. After the fiasco in Massachusetts, where clearly he had been advocating things for which the time and place were not ready, he settled down in less risky activities. Now planning to settle down in an Irish academy, now writing an essay on the "Defects and Remedies of

[49] R. P. Stearns, *op. cit.*, Essex Institute, *Hist. Coll.*, LXII (1936), 54–56.
[50] W. Orme, ed., *Remarkable Passages in the Life of William Kiffen Written by Himself* (London and Perth, 1823), 33–36.

English Husbandry," he wound up finally as an agricultural expert for a large holder of Irish lands.

While his financial stake in the ironworks came to £450 by 1647,[51] Child differed from other large investors in that he was to be on the scene, and with a store of technical knowledge which certainly matched and probably excelled that of Winthrop, the original promoter. He had tried to help in the taking on of additional workers and materials after Winthrop's departure. He had answered Winthrop's metallurgical questions. Like Downing, he had kept him informed of the developing plans of the Undertakers. For part of his stay in Massachusetts, he seems actually to have worked at the ironworks. His surviving writings indicate a clear, practical approach to scientific and industrial problems. Had he not become embroiled in the politics of the Great Remonstrance, he might have been, the pressures of his multiple interests notwithstanding, the perfect link between the English firm and its American operations.

Such, then, were the men who risked their money that New England might have an ironworks. Differing in their professional backgrounds but having many business contacts, differing in the amount of the world's goods at their command, they were typical of the capitalists of their day, and in many respects typical of those of modern times. They made their investments with great expectation of making profits. The sums involved were high. The initial capital subscription came to £1000, probably in full shares of £100. By 1651 some of the shareholders had advanced £650 for each £100 of initial subscription.[52] By 1653 at least £12,000 had been invested in the enterprise.[53] However great the expectations, however encouraging the inducements offered by the Massachusetts authorities, investment of such proportions took courage. The general political situation was full of uncertainties. The Company organization and its techniques for control and supervision of the investment were loose and even crude, as will be more than apparent in the record of actual operations. Though these men remained in England, they qualify, through the medium of their venture capital, among those whom we are accustomed to call pioneers of American development.

[51] *Mass. Records*, II, 450.   [52] Suffolk R. & F., 125–2.
[53] Mass. Arch., 59/53. The Undertakers cited an even higher figure, £15,000, *circa* 1661. *(C. S. P. Col.*, America and West Indies, 1661–68, p. 17.)

Not all of them stayed in England. Three of them, as we have seen, emigrated and remained in Massachusetts for shorter or longer periods. In additon to the English capitalists and to these transients, there were certain permanent settlers in the Puritan colony who eventually invested money in Winthrop's ironworks. Local investment was strongly encouraged by the General Court. It would have been strange indeed, if Winthrop's promotional talents had not borne some fruit when he reached home with glowing reports of rich home-country support of his undertaking. It is easy to imagine him offering some of his colleagues a chance to get in on the ground floor of what looked like a fine investment opportunity. From all we can learn, however, there was no rush to get aboard his bandwagon.[54] We can tell little of the nature and extent of the holdings of local residents. One investor, Henry Webb, has already been mentioned. The limited data available on others are best postponed until we pick up the story of actual ironworks operations. It is enough to indicate here that while the Company of Undertakers was predominantly an English group, its efforts had the support of at least a few settled colonists as well as that of the Massachusetts authorities.

The motivation for the investment of large sums by these men was simple. They were predominantly Puritans, and they wanted to help in the development of the Puritan settlement in Massachusetts. They wanted to make money. The former is rarely explicit in surviving documents. The latter is clear in a dozen deeds, grants, and letters. However great Winthrop's talents as a promoter, and however strong the capitalists' Puritan sympathies, it is more than doubtful if men of the calibre and degree of business experience of most of the Undertakers would have put in their money had there not been good reason to believe that a New England ironworks would make money.

Many of the shareholders had had ample opportunity to learn of the amazing record of growth, by immigration and natural increase, of the Massachusetts Bay group during the 1630's. Some of them knew, as well as any men could, the impact of political upheaval and civil war on a remote colony hitherto dependent on regular importations from the mother country. We know, however, that the hopes of these investors were pinned not alone on the prospects of profits from

[54] According to Winthrop's *History of New England,* "only some two or three private persons" joined in the venture. (Savage, ed., II, 261).

meeting the colonists' needs for iron. They looked forward from the beginning to the day when New England ironworks would provide iron for export to England. This was to be, in other words, a case of economic imperialism. English capital was being exported to produce a basic commodity for the Mother Country's use. Given the basic conditions then prevalent in the English iron trade, such a venture made sense and promised much.

Thanks largely to pressures of increased demands on an industry whose charcoal timber resources were being pushed to the limit, there was money to be made in iron. Ample evidence is available to document what the classical theory of supply and demand has taught us. In the first place we find that the price of iron rose markedly in the period 1620–1660. The range of prices in the early years may be seen in the following table:

The Price of Forest (Bar) Iron at Chatham[55]

Year	Minimum Per Ton			Maximum Per Ton		
	£	s.	d.	£	s.	d.
1619	13	17	9	14	10	0
1620	14	10	0	15	5	0
1621	14	0	0	14	10	0
1622	13	6	8	14	0	0
1623	13	10	0	13	10	0
1625	14	5	0	14	5	0
1630	17	18	0	17	18	0

The slump in the early twenties was ascribed to a general shortage of money, product of investment in the East India Company, the Virginia lotteries, and the costs of the Bohemian wars.[56] It was clearly temporary, however, as is clear in both the table and in the Day Book of the Burrells, where the price of iron ran from £15.10s. a ton to £17.2s.6d. between 1636 and 1656.[57] In another quarter, Sir E. Osborne reported to the Privy Council in favor of allowing Lionel

[55] James E. Thorold Rogers, *A History of Agriculture and Prices in England* (7 vols. in 8, Oxford, 1866-1902), VI, 449 *et seq*.

[56] Richard Boyle, 1st Earl of Cork, *The Lismore Papers (Second Series)* (A. B. Grosart, ed., 5 vols., London, 1887–88), II, 244–46, 263.

[57] *Sussex Arch. Coll.*, XLIII (1900), 14.

Copley and his partners to build an ironworks at Conisborough, in Yorkshire, in 1639, and said, "I am of opinion that the erecting of that work and felling of wood can be no prejudice to the public but rather a benefit in lessening the price of iron which is sold at an extreme high rate."[58] The extreme high rate doubtless prevailed when Winthrop reached England in his quest for capital. In fact, it seems to have persisted for another twenty years. Bar iron was sold from the Commonwealth's ironworks in the Forest of Dean in 1657 at prices from £17.10s. to £18. Pig iron prices also increased over the period in question from £5 in 1620 to £6.6s.8d.–£7.5s. in 1653–1656, and to £7–£7.7s.6d. in 1657.[59]

Secondly, it is a commonplace of economic theory that profit margins grow wider in periods of rising prices. In the history of the iron industry we are not without evidence of an unusually large gap between the cost of producing iron and the prices it fetched. This gap resulted in the development of the industry in areas which at first glance seem most unpromising. The most conspicuous examples of the trend are to be found in Ireland. Readily apparent obstacles, lack of skilled workers, distance from the English market, even absence of iron ore, did not deter investors from building what must have been high-cost plants. The Earl of Cork imported workers and at times imported ore—and was reported to have made 40 per cent profit per year, with cost of production and shipment given at one point as £10–£11 and sale price as £16–£17 per ton.[60] In one instance it was apparently profitable even to scour the Low Countries for workers for the Irish ironworks.[61]

The English home industry was also good business, as the career of Thomas Foley, the Undertaker, would indicate, even if Baxter's claim of a rise from almost nothing to £5000 a year from his ironworks is exaggerated. For other evidence, there is Lionel Copley's

---

[58] *C. S. P. D. Charles I*, 1638–39, pp. 516–17.

[59] S. P. 18/157B; S. P. 18/130/102.

[60] Gerard Boate, *Irelands Naturall History* (London, 1652), in *A Collection of Tracts and Treatises Illustrative of the Natural History, Antiquities and the Political and Social State of Ireland* (2 vols., Dublin, 1860–61), I, 111–12.

[61] In 1633 Richard Rowley and Jacques Lagasse went to "Lukeland in the parte beyond the Seas" to recruit skilled hands for the Irish ironworks of Joshua Foote, William Beeke and their partners, and secured 16 Walloons. (Rawlinson MSS., D 918/133.)

1639 lease of some land near Sheffield belonging to the Earl of Arundel. Copley and his partners were to pay £2120.10s. a year for twenty-one years. In return they were allowed the use of the land, the cutting of seven thousand loads of wood per year, the right to dig for ironstone, and liberty to erect an ironworks.[62] In a profusion of surviving legal papers and petitions, there is no mention of any compensation to Copley and his group for their capital outlay when the lease expired. It would appear that they hoped within the given period to recover their capital and make a profit as well! This conjecture is supported by an offer that George Sitwell made in 1664 to the Marquis of Newcastle. He wrote that if he might have the lease of the necessary land and watercourse, he would build at his own expense a forge, workmen's houses, dams, and the like, pay two years' rent as a fine, and leave all standing at the end of the twenty-one-year-lease term.[63]

A growing population in the Mother Country cried out for more and more iron. A charcoal timber shortage had pushed the English iron industry from its ancient centers to ever more remote areas. It had been carried to Ireland and to Scotland, where in certain cases even the ore had to be imported from remote places and much of the finished and weighty product carried back to English markets. The reaching out of far-sighted capitalists to New England thus seems to be little more than an extension of an already well-established trend of economic imperialism fed by the lure of high profits in a generally favorable business situation. On this side of the Atlantic were virgin timber resources of staggering extent. Here, Winthrop had assured the capitalists, were ample stores of iron ore, a market for some of the product, tender concern from civil authorities for a much-needed infant industry. If our analysis of the state of the English iron industry is correct, Winthrop could not have picked a better time to launch his capital-raising venture. It is difficult to see how he could have failed in it.

[62] C5/22/27 and S. P. 23/115/1005.
[63] Derbyshire Archaeological and Natural History Society, *Journal*, X (1888), 35.

5

Government and the
Infant Iron Industry

THE NEW ENGLAND IRONWORKS were the products of public as well as private enterprise. However auspicious the general economic situation, it is more than doubtful if the Undertakers would have chosen to risk their money had they not been assured of special concessions and privileges. True, the plants erected in Scotland and Ireland had been straight private-risk ventures. These, however, were a far cry from heavy fixed investment in a remote semiwilderness in a period of grave political uncertainties. Even if they had concluded, despite these drawbacks, that New England iron was a promising investment opportunity, it would have been inconceivable for the magistrates to follow a hands-off policy with the fine new works which the Undertakers' money was to set up in their colony. The prevailing climate of opinion guaranteed that if the establishment of the ironworks was a function of capital investment by a group of entrepreneurs, it had also to be a function of government assistance *and* government regulation.

The Puritan community in Massachusetts was in many respects a special case. The essentially medieval overtones of the "organic state," in this instance a society of true believers gathered together to serve God according to one "right" way, still prevailed. Every aspect of life had to square with the dominant orientation of the whole group as interpreted by its leaders, lay and spiritual. Worthy economic activities were encouraged. Those which were deemed otherwise were forbidden. If exhortation were not sufficient to persuade toward, then regulation was to enforce, the "conformity" which was the most precious entity in the Puritan *Weltanschauung*, as Robert Child had learned to his pain. Intervention could, and actually did, come close to covering every phase of life, religious, social, intellectual, and economic. It is not that there was no freedom. Rather, each freedom was balanced by a specific complement of responsibility to the community at large, and to God. Since man was weak and depraved in consequence of Adam's fall, the arm of the state stood ready to see the responsibility enforced whenever necessary. *Laissez faire* was unheard of.

Seventeenth-century Massachusetts was at least an approximation of the organic state. It was also a new and, in some degree, a planned community in which the policies we lump under the general heading of mercantilism loomed almost as large as the religiously inspired elements just cited. Mercantilism may be loosely defined as the regulation, by government, of a country's economic activities in the light of what is held to be the nation's best economic interest. The policies involved range from fiscal controls and the regulation of external trade to the stimulation of new industries and the supervision of established ones. The prime emphasis in the Bay Colony fell, of course, on the development of new industries which would safeguard its position if imports were cut off by war or by shifts in the balance of trade. We have already seen how willing the authorities had been to encourage even the crudest beginnings in certain fields, especially under the impact of the depression which followed the slump in immigration and importation.

Few materials were as vital for the successful development of the Colony as iron, and great indeed was the extent to which the magistrates were pleased to go in making possible its effective production within their jurisdiction. Since their encouragement amounted to government subsidy, it was welcomed with open arms by the capitalists. The more they could get from the General Court, the better, and we shall soon see that they were anything but laggard in their demands. What Winthrop had been able to take with him to England as assurance of government support of an infant industry was only the beginning.

The encouragement, however, had a complement of regulation. In Massachusetts, the general urge to regulate was strong enough. Its leaders were also aware of the lessons of old-country experience with government assistance to industry. These would clearly have given an additional push to an inclination which was part and parcel of the arrangements of an organic state. Government aid to industry had come to the fore in England during the reign of Elizabeth, and the English people had had ample opportunity to see that, while capitalists were always eager for monopolies and patents, they could not always be trusted to use well the special privileges which governments, for one reason or another, had granted them. If the General Court chose to offer much by way of encouragement, it would also have made sure to try so to tie the capitalists' hands that all would benefit, and none, least of all the general community, be harmed. On this much, at least, mercantilism and practical experience with mercantilism were in accord.

When the public authorities of Massachusetts granted monopolies on the manufacture of salt, iron, and the like, they were doing what had been done in England for many decades, and on a more extensive scale than later generations, steeped in the *laissez-faire* tradition, have often chosen to remember. Some forty-five years ago, Hermann Levy of Heidelberg pointed out that "industrial capitalism in England was cradled in monopoly, not in competition."[1] The monopoly to which he referred was not normally the product of elimination of competition by business practices. It was a *legal* monopoly, often on a national scale, resting in a grant from the Crown of exclusive rights of manufacture and often accompanied by techniques of *legal* suppression of competition.

These grants from the Crown were, at least in theory, mercantilist in inspiration. Originally, they were little different from the patents with which modern nations protect inventors and the capitalists who subsidize the practical development of their inventions. In 1567, for example, a group of workers petitioned for exclusive rights to make window glass for twenty-one years. After checking to make sure that this would work no injury on the established glass industry, the

[1] Hermann Levy, *Monopoly and Competition, a Study in English Industrial Organisation* (London, 1911), 42.

Crown granted the appropriate "patent."[2] One for the making of salt had been awarded the year before. In time there were others which covered white soap, saltpetre, alum, oils, and similar items which, in the past, England had had to import from the Continent. In all these the government was simply encouraging infant industries by providing patent protection to capitalists who proposed to develop new or newly imported production techniques.

During Elizabeth's reign, however, the patents or monopolies multiplied enormously, the bases of award were less strictly focused on novelty or usefulness to the country, and monopolies in trade fell increasingly into the hands of court favorites and others whose interests were selfish and whose methods often unscrupulous. By 1601 popular protests reached such heights that many of the monopolies were canceled, with Elizabeth insisting that she had never granted one which she considered *malum in se,* and thanking the Commons for calling the abuses to her attention. Monopolies continued to be granted under James I, however, and presumably reached their peak between 1614 and 1621.[3] Continuing public pressures led to the Anti-Monopoly Act of 1624, but this was less than effective. It left several monopoly areas untouched. It also left the door open to others in its provision of fourteen-year privileges on new inventions. In 1639, Charles I promised to restrict the monopolies, explaining that the privileges had been awarded on the pretense that they would benefit his subjects' "common good and profit," but had turned out to be "prejudicial and inconvenient to the people."[4] Many monopolies were canceled after 1640, but it was not until 1689 and the repeal of the royal prerogative that the very legal foundations of monopoly were eliminated. In effect, the whole period from 1600 to 1685 was rich in monopolies—and in protests arising from the discrepancies between the benefits to the realm alleged by those who sought them and their concrete results.

A few examples of the actual working out of some of these monopolies may shed light on why the magistrates acted as they did in their dealings with the Company of Undertakers. The manufacture of salt

[2] W. Cunningham, *The Growth of English Industry and Commerce in Modern Times* (2 vols. in 3, Cambridge, 1927–38), II, 76.

[3] Levy, *op. cit.,* 17, 46.

[4] *Ibid.,* 46, 67.

in England enjoyed monopoly privileges from 1565 to 1601. By the latter date its price had increased tenfold and Elizabeth withdrew the privileges. James I chose not to renew it. For a while, war and consequent limitation of imports favored the English producers, but when normalcy returned they again sought legal protection. From Charles I they obtained the right to set up a corporation, the Company of Saltmakers of North and South Shields, with a monopoly of salt manufacture on the coast from Southampton to Newcastle. Scottish producers were ordered to join with it. It was also favored with a ban on importation of salt from the Continent. It turned out salt, but its price more than doubled in the period between 1630–1635 and 1640, and was much higher than that prevailing in the non-monopoly areas of England.

Similar developments arose in the case of a monopoly covering the making of drinking glasses. In 1574 an Italian got a basic patent which, in 1615, fell into the hands of Sir Robert Mansell, capitalist and courtier. The furnaces involved were fired by coal. Since Mansell's competitors used wood, and wood was increasingly scarce, it took only a ban on the firing of glass with wood, and restrictions on importation, to give him what amounted to a monopoly of all the glass manufacture of England. Untouched by the Anti-Monopoly Act of 1624, and well supported by the Privy Council in efforts to suppress would-be competitors, Mansell managed to maintain his nice position until 1642, when the monopoly was finally canceled. In the interim, it had led to high prices and poor goods.[5]

Monopolies in the mineral field were more complicated. In alum, whose production was new to England, monopoly and prohibition of imports prevailed from 1607 to 1648. Thus securely protected, capitalists invested more than £20,000 in a short period, in highly inferior deposits in Yorkshire. Despite continuing capital outlay, they never managed to meet production costs, and fell far short of meeting English needs. The result was widespread protests over high prices and low quality. In the case of tin, certainly no new industry, rising costs as mines went deeper persuaded Elizabeth to hand over her rights in the Cornish mines to a group of capitalists. From this stemmed an industry-wide monopoly of the London Pewterers, who, in order to control one of their vital raw materials, bought up the

<hr />

5 *Ibid.*, 33–34, 47.

royal privilege in 1615. They maintained their privileged position for some years, to the great hardship of the tinners of England and, one assumes, the English people. Similarly, Elizabeth's rights in the coal mines of Newcastle went first to the town, thence to a small group of merchant coal miners who were set up in 1600 with a monopoly of the better mines and exclusive export rights. Surviving the Anti-Monopoly Act, the group held its privileges, not without frequent challenge, until after 1653. Production was controlled, quality hinged on no more than the monopolists' consciences, and the natural result was that the people at large suffered.[6]

The English ironworks had been established without benefit of monopolies, and their regulation was almost wholly a matter of fuel conservation. Anything which promised to reduce the quantities of wood that went, in the form of charcoal, into the ironmasters' furnaces and forges was, of course, highly worthy of government support. The list of privileges allowed on the use of coal in smelting is long. Twenty-three of the hundred-odd patents granted between 1620 and 1640 had, in one way or another, to do with furnaces, ovens, smelting, and refining.[7] Not even this old and well-established industry was wholly free from monopolies of the kind in which personal or political privilege figured larger than special inducement to industrial development. In 1612, for example, the Earl of Pembroke secured exclusive rights to take coal and iron from the Forest of Dean, although only after a working compromise with the interests of the so-called free miners, the people of the ancient mining community of the area, had been achieved.[8] Obviously, however, precedents for the policies pursued by the Massachusetts magistrates are to be found, not in the English iron industry, but in mineral fields in which the authorities were bent on stimulating initial industrial development.

One such is the work of the Society of Mines Royal, by which England hoped to render herself independent of the importation of copper. In 1564, Houghstetter and Thurland obtained a patent for the exclusive right to dig for gold, silver, copper, and quicksilver in the northern and western counties of England and in Wales. The

<hr/>

[6] *Ibid.*, 7–8, 24–30.

[7] William Hyde Price, *The English Patents of Monopoly* (Boston and New York, 1906), 111 n.

[8] Levy, op. cit., 30–31.

privileges awarded then and thereafter were much more liberal than, but also had much resemblance to, those which Massachusetts awarded the Undertakers. Among them were authorization to purchase land and hire workmen at reasonable rates, sole rights to the use of any tools not used in England for twenty years, exemption from subsidies and fifteenths, the privilege of using the Queen's timber for fuel and construction, and permission to take their own care of disorderly persons in their employ and to keep a tavern at their works. In 1568 they received a charter of incorporation. Although their operations, which were mainly in copper, were not successful, they secured another charter in the second year of the reign of James I, by which the Society continued to be vested with a monopoly of metal mining in northern England and Wales. Eventually, it became a mere privilege monger, granting leases to real entrepreneurs, some of whom were at least moderately successful.[9]

Another, and more or less parallel, venture was the Mineral and Battery Works, a company endowed with similar privileges in the eastern and southern counties of England. Humphrey and Shutz obtained two patents in 1565, one for gold, silver, copper, and quicksilver, the other for calamine stone, an ore containing zinc, then much in demand for use with copper in the making of latten. Three years later they and their backers were incorporated. Their special privileges even applied to *future* inventions in their field. The results obtained from the lavish encouragement granted them were not impressive. At Tintern, in Monmouthshire, the group introduced mechanical wire drawing by the standard technique of importing foreign workmen. While much money was poured into the enterprise, and while its sponsors were protected by the courts against infringers, and against foreign competition by a Parliamentary ban on imports of wool cards, for which most of the wire was destined, the operation was not successful. Activity with the calamine stone was not started for twenty years, and when it was, it was a complete failure. As with its sister company, effective work was to come only from the efforts of individuals operating under licenses from the monopolists.[10]

In both cases the government had acted on the assumption that to initiate work on untapped mineral resources requiring both the

[9] *Ibid.,* 49–55.
[10] *Ibid.,* 55–61.

importation of techniques and skilled workmen and large and long-continued investment, it made good sense to offer sweeping inducements to capitalists. The monopolies granted to the Mines Royal and to the Mineral and Battery Works were too broad and too rigid. They led to abuses as serious as those which arose from some of the more patently political grants of privileges mentioned earlier. The industries thus falteringly introduced to England were, however, eventually firmly established. The monopolies did get the needed specialists imported; they naturalized certain of the metallurgical arts. More important still, they provided a focus and a frame for the investment and employment of capital.

These were, of course, the first important broad monopolies granted in England. Further experience made the authorities more cautious, except where favoritism and politics outweighed common sense and sound mercantilism. When the Massachusetts officials found themselves in much the same situation with respect to iron that England had been in with regard to copper and zinc, it was not strange that they should adopt a policy of generous encouragement. The record of the home country's total experience in this business of government assistance to industry would certainly have suggested that they be circumspect as they did so. A history with which they must have been familiar reinforced a belief which they had taken from John Calvin. Capitalists and entrepreneurs, like all men, still bore traces of the taint inherited from Adam's sin. If it took assurance of special privileges to get the necessary capital investment, let it be forthcoming. Since capitalists were capitalists, however, let the privileges be tightly drawn, the interests of the whole community protected by adequate regulation, especially of the prices which the new infant industry could charge a people much in need of its products.

While the resolution of the General Court under the heading of "Encouragement to mines, etc.," which Winthrop carried to England, did not mention the word "monopoly," it did offer full control of mineral operations at particular sites for twenty-one years. The colony treasury was not to receive any fraction of the proceeds of the type which had historically been part of the rights of sovereignty in the case of precious metals. There were also other less important guarantees. All this would hardly have been sufficient, however, to suit English capitalists familiar with far more lavish grants in support of new industries. It is probable that the Undertakers coupled a demand that Winthrop try to persuade the Court to do better by them with their offer of £1000 for initial capitalization. Certainly, Winthrop must have made application for extension and clarification of the Court's "encouragement" during the winter of 1643–1644, since in the spring the magistrates passed a measure which was far more generous, more like some of the English grants mentioned above. In the form in which we have it, it appears as "answers" to "propositions" presumably submitted by Winthrop, and since lost.

By its terms the Undertakers of New England ironworks and those who joined them by investment of £100 before the month's end were to receive a twenty-one-year monopoly of iron manufacture within the jurisdiction of Massachusetts, grants of common land of unspecified area, access to private property, with the owners' consent, in a form to be decided on later, rights to export iron in excess of the Colony's requirements, and ten years' freedom from taxes. Ironworks could be established, and presumably with separate land grants, in as many as six places. To get the monopoly, the ironworks had to produce enough iron for the Colony's use within two years. To get the land at the various sites, the Undertakers had to build "an iron furnace (and) forge" in each of the six places and not "a blomery onely." General privileges were to be only "such as the lawes of the country doth alow." While these last items clearly indicate a policy of circumspection, it is interesting that at this stage the magistrates had not seen fit to set ceiling prices on the ironworks products.[11]

These "answers" date from March, 1644. Apparently, they were still insufficient to please the Undertakers. In November, the Court produced another set of answers to propositions which it had received from the English investors, presumably through young Winthrop, in the interval. Here the grants and privileges are better spelled out and in some cases extended. So, however, are the magistrates' restrictions. Three years from the date of the answers were allowed for the works' completion. The investment subscription list was to be open to Massachusetts residents for a full year. The size of the land grant for each of the six plant locations was set at three square miles. It was to hold in perpetuity. To the Undertakers went rights

[11] *Mass Records,* I, 327; II, 61–62; *2 M. H. S. Proc.,* VIII, 15; *Winthrop Papers,* IV, 450–51.

to ore, water power, timber and building materials in common lands, and liberty for highways and watercourses in private property. Any other "accommodations" in private property, however, had to rest on a vaguely expressed willingness of the inhabitants to extend them by composition or otherwise. These presumably involved ore and timber taking. The Company and its servants were assured full equality before the law. It also got a twenty-year tax exemption on all stock and goods employed in the ironworks. Its employees were dispensed from militia service. On the other side of the ledger, that of restrictions, the Court asked the Company to provide some form of religious instruction at plants erected far from established churches. It also set £20 as the maximum sale price on bar iron.[12]

The negotiations to this point had doubtless been carried by John Winthrop the Younger. Their results were still less than satisfactory to the gentlemen investors back home. For reasons to be discussed below, but among which displeasure over the fruits of Winthrop's efforts probably played some part, he was replaced as agent by Richard Leader in the summer of 1645. Getting further clarification of the various rights and privileges, and the confirmation of those rights and privileges in a formal charter, was probably the first chore of the Undertakers' new representative.

With him, we know, came an eloquent if disingenuous letter from the English group, which announced that it felt "utterly disabled to carry on the Worke to the publique Advantage," unless the magistrates reacted favorably to a plea for further privileges, the details and justification of which Leader was to deliver to them. The Undertakers claimed that the "charge & Hazard" of ironworks was to be all theirs, the "Advantage" wholly that of the magistrates and the general public. They insisted that they found themselves "much straitned in diverse of the particulars" of the encouragement offered to date. Let them have what they wanted, and with God's help New England would have its ironworks. Great public benefit was assured. So were "good Returnes" to all of the colonists who would help in the project. At no point, however, was there even a suggestion that the Undertakers hoped to make money with their venture![13]

The product of the new negotiations was the formal grant or

12 *Mass. Records,* II, 81–82.
13 Mass. Arch., 59/13.

charter voted by the General Court on October 1, 1645. Its contents are little different from the concessions, privileges, and qualifications spelled out in the answers to the earlier propositions. There was, however, a certain amount of tightening up of the various arrangements. If the Undertakers and any who chose to join them, this time by investment of £50 by the end of October, made enough iron to meet the Colony's needs within three years from the date of the former grant, and sold it for no more than £20 a ton, they would receive a monopoly of the manufacture of iron, and exclusive rights in all iron mines, discovered and undiscovered, within the limits of Massachusetts for twenty-one years. Their rights to take ore and charcoal timber, building materials and the like, and to build ways and watercourses in common land, likewise extending over twenty-one years, were unlimited except for responsibility for the flooding of private property. On land in private ownership the Company and its servants could take out ore and build and use ways and watercourses, except where houses, yards, and small orchards were involved, by paying the owners sums agreed upon as fair by three umpires chosen by the Court, the Company, and the property owner, respectively.

Outright land grants were again authorized in as many as six places. Each was to cover three square miles, but no dimension was to exceed four miles, and each ironworks was to consist of furnace and forges, not just a bloomery, and had to be completed within a term of years left blank in the original document. Should conflict arise between the Colony's plans for settlement and the Company's plant-building projects, the latter were to have priority, the Court pledging due advance notice and assuring the Undertakers "first choice." Export rights were restricted only to the extent that local needs had first to be met, and goods could not be sold to persons or states "in actual hostility" with Massachusetts. Equal rights before the law were granted in perpetuity. Exemption from taxes and all public charges whatsoever was to hold good for a period left unspecified in the grant. The term of exemption of full-time workers from military service was also left open, although the magistrates now insisted that they at least be provided with arms and ammunition. Again there was the stipulation that religous instruction be provided in places remote from churches or settled congregations.[14]

[14] *Mass. Records*, II, 125–28.

The English investors wanted still more. In May, 1646, they wrote another letter to the magistrates. It has not survived, but some of its contents may be inferred from the reply it produced from the General Court. Entered in its records under date of November 4, 1646, this was presumably the product of the deliberations of a special committee appointed to ponder the Undertakers' letter and make return of "w^t they conceave meete to be done thereabouts." It began with a not too gentle bit of knuckle rapping. The magistrates made it clear that they found the style of the Company's letter more sharp, and its conclusions more peremptory, than rational. They excused this, saying "wee consid^r yew have binn hitherto loose^{r}s, & therefore may take leave to speake." Much of the letter concerned the Colony's money shortage, and it may be that the Undertakers had asked to have the ceiling price eliminated or restricted to transactions in hard money. Clearly they had asked for the full twenty-one years to erect plants and collect the appropriate land grants. This the Court looked on with disfavor. It might well work a hardship on people inclined to build ironworks after the monopoly expired, especially if the Undertakers had all the best sites tied up. The general tone of the letter is that of patient explanation of the Court's position. It concluded with an expression of the magistrates' hope that future differences might be avoided and an "aequall" way found by which the colonists' needs could be met and the Undertakers encouraged to proceed with their project.[15]

In the same month the Court came through with still another set of answers to propositions from the Undertakers. Here the Company chalked up impressive gains. The twenty-one-year monopoly was to run from October 1, 1645, rather than from March 7, 1644. If the Court saw fit not to allow the full period for plant completion, it seems to have granted ten years. It also announced that while the Undertakers might build as many as six ironworks, they could have three square miles per finished plant even though they were not at different sites. In theory, at least, the erection of six furnaces and their accompanying forges at one place would have given the Company an industrial empire of eighteen square miles. In answer to misgivings about the prohibition of sales to enemies, the Court assured the Undertakers that here they were being treated no differ-

15 *Ibid.*, III, 80, 91–93.

ently from the regular inhabitants. The Company had also apparently asked to have all in its employ free from military service. The Court again restricted the privilege to regular, full-time employees. It did, however, free the Company of the need to provide religious instruction, a major concession indeed. It also acceded to its request that the blanks in the grant voted the previous year be filled in, although we are not told how.[16]

The final product of all these negotiations was engrossed on parchment over the Court's seal and sent to England. It had not been received by March, 1647, but Leader had provided his employers with a transcript. Then, at long, long last, the Undertakers announced that they were content. Both the formal grant and the copy have been lost, but one would infer that the Undertakers had reason to be satisfied. Now, however, they wrote the magistrates, not to express thanks, but to point out that they had been asking only for the minimum needed to make a success of the venture. If the privileges seemed large in Massachusetts eyes, they had not been sufficient, they wrote, to persuade some of the English capitalists to persevere in their investment. And those who stayed were now but quietly waiting for a harvest not yet in blade. They had good reason to continue in the magistrates' good graces, however, and therefore announced that they had rechecked the contents of their letter of demands to see why they had been called to task for "sharpness of style, peremtoriness and want of rationality." They insisted that they knew not wherein they had offended but added that they were but men and prone to error. Surely, they said, "He is more than a man that doth not do so sometimes, and less a man than he ought to be that is not willing to be admonished and to amend."

An approximation of an apology offered, the Undertakers then had the audacity to interfere in the unfortunate case of Doctor Child. He had been charged with verbal instructions; because of his incarceration, the Company's business would suffer. They admitted that they were in no position to judge the merits of the case. They professed that they were not meddling. They nevertheless suggested that the magistrates ascribe his crime to "common frailty" and let him go free with an admonition. In what appears to be only a thinly veiled threat, they coupled their request with a suggestion that if it were granted

16 *Ibid.*, II, 185–86.

they would not "engage in anything that might interrupt your power."[17] This can only refer to the possibility of some kind of proceedings in the Mother Country against the Colony's charter rights. Those who have found too much government "interference" with the ironworks business in the above summary of negotiations between the Company and the Massachusetts authorities must admit that here was a clear case of business interference with government.

The encouragement finally voted the Company of Undertakers was by all counts lavish. In unsettled areas it got extensive land grants and free access to raw materials. In securing what it needed from the holders of private property it was at least protected from any tendency of theirs to "hold up" what was, for the time, big business. Its monopoly of iron manufacture was absolute if it could meet the Colony's needs. The £20-price-ceiling should have allowed an adequate margin of protection from the competition of the home industry, whose product was selling for less than £18.[18] The tax exemption was at least a minor indirect subsidy. So were the exemptions of regular employees from militia service and the exemption of the Company from what, in wilderness areas, might have been the costly necessity of maintaining a minister. About the only thing in the whole book of government assistance to industry in the seventeenth century which the Undertakers did not receive were outright bounties on production and the prohibition of iron importation. It is doubtful if any single private enterprise in New England, in the whole colonial period, was so well favored by government.

The Massachusetts authorities fell short of complying with the capitalists' demands. As the negotiations proceeded, however, it is clear that they were driven to further and further concessions. In their final form, it may be assumed that the grants, privileges, and immunities far outweighed the restrictions imposed in the interests of the whole community. Effective iron production would be of enormous benefit to the Colony, and no one knew it better than the magistrates. If the long-drawn-out negotiations are regarded as a contest between business and government, then it appears that business had won. The facts of the New England economy were on its side.

The tug of war between business and government was not over,

17 Suffolk R. & F., 125.
18 *Sussex Arch. Coll.,* XLIII (1900), 14.

however, when these magnificent privileges were confirmed by formal charter. In the autumn of 1648 the inhabitants of Lynn petitioned the General Court for clarification of the Company's tax exemption. They asked if it applied to the taxes of particular towns and churches. The Court's answer went against the Company, the magistrates claiming that "by publicke rates, taxes, &c, are ment rates, leuies, or assessments of the common wealth, & not of the towne or church."[19] The Company of course protested; indeed, the Undertakers were filing their objections on this point as late as August, 1650. A crude draft of the Court's response has survived, and helps round out the picture of relations between government and industry. In it the Court claimed that it lacked the power to grant exemption from local taxes, and pointed out that the present problem would never have arisen had the Company's agent voluntarily contributed to the support of Lynn town and church. Despite this, it had prevailed on the town to keep its annual charges on the ironworks under £20.

In this document we find evidence that the magistrates had some ammunition with which to back up their side. They wrote:

We haue always valued the love & good will of o^r friends att a high rate & would not willingly forfeit o^r interest in any such much lesse in y^{r}selves were wee not insenseable of some iniurye done vs (as wee are credibly informed) by some that hath relacon to the Ironwork, but that God whom wee serve & whoe hitherto hath bene o^r guide, will in his tyme cleare o^r iniury & wype of such aspercons as formerly hee hath done, blessed be his name.

We cannot tell what injury had been involved. Even here, however, the Company obtained the ultimate victory. Scrawled at the bottom of an appended page was a note that, if so small a matter was to cause difference and disaffection between the Undertakers and the Colony, the latter would itself pay the relatively light sum involved.[20] Eventually, it did just this. In May, 1653, the Court granted Lynn £10 per year in lieu of the taxes which the town should have got from the ironworks. Apparently, the English gentlemen had successfully maintained that exemption from "all taxes, assessments, contributions, & other publiq charges whatsoever" meant just what it said.[21]

19 *Mass. Records*, III, 142.
20 Mass. Arch., 59/40.
21 *Mass. Records*, III, 307; IV–1, 137; V, 127.

Within a few months of the voting of the special subsidy to Lynn, the financial difficulties of the ironworks came to a peak, and various creditors took over its assets on order of various courts in which they had lodged actions for debt. The Undertakers ultimately came to feel that they had been victims of gross miscarriage of justice, that they had been robbed. We shall see, however, that even as the assets were being handed to the creditors, the Massachusetts authorities were insisting that they were acting in the best interests of the Company, all appearances to the contrary notwithstanding. There is still another instance of the General Court's solicitude for its obligations to the Company of Undertakers. In the fall of 1657 the Court granted encouragement, and, incidentally, on notably less lavish terms than had prevailed for the Undertakers, for ironworks in the vicinity of Concord and Lancaster. This was, of course, a violation of the monopoly enjoyed by the Undertakers and, one assumes, of the present holders of their property. It was justified on the grounds that their plant was likely to cease operations, that often no iron was available, and that at times it had been sold for more than the ceiling price when it was available. In summary, the obligation to meet the Colony's need for iron had not been met.

Despite these presumably quite adequate reasons, the Court bent over backwards to be fair to the Undertakers. It had asked the present holders of the ironworks property if they would take on the commitment to satisfy Massachusetts' needs. Receiving no such assurance, it was authorizing the setting up of new works. These, however, might be bought up by the Undertakers at a price to be set by neutral umpires, if they felt so inclined. They were guaranteed more than a year in which to exercise this privilege. Further, the proprietors of the new ironworks were not to employ any of the Company's workmen without its consent, "that so the setting vp of one worke may not pjudice the old."[22] Certainly, the Undertakers' interests were not getting the brush-off that their record of performance to date might seem to have abundantly justified.

Besides the assistance from the commonwealth, the Undertakers also received generous help from town governments. Winthrop, early in the game, got extensive land grants from the towns of Boston and Dorchester. The motivation for these was presumably little dif-

[22] *Ibid.*, IV–1, 311–12.

ferent from that of today's local chambers of commerce working to attract promising new enterprises to their areas. In all likelihood, there was a similar grant from the common lands of Lynn, although surviving records do not make clear either the fact or the areas that may have been involved.[23] Since the Company was able to erect a complete ironworks at Lynn within the specified period, it should have gotten its three square miles, whether by General Court or local town government action. In time, relations between local jurisdictions and the Company became increasingly strained. At Lynn, for example, the problems posed by workers' lax—or different—morals became far more serious than the wrangle over taxation just mentioned. In the early years of the Company's operations, however, the local authorities had been as generous as the General Court in offering concrete encouragement to the infant iron industry.

If New England's first ironworks were to end in a business failure, it was not because they were put through in an unfavorable atmosphere. From the point of view of the Puritan leaders, this was to be large-scale combined and complementary effort toward the goal of iron to meet a country's needs and profits for those who took the risk involved in making it possible. The investors, if only from their Puritan sympathies, had at least some degree of service motivation. The Puritan magistrates would have been the last people in the world to question a business man's right to his fair profit. If industrial enterprise in England was cradled in monopoly, so were the effective beginnings of the American iron industry. Here was a joining of profit motive and policies appropriate to the Puritan organic state and to mercantilism that should, in theory, have permitted the infant iron industry to wax robust and vigorous.

[23] A 1684 deed conveying old rights in the ironworks property at Braintree gives a kind of historical summary of the launching of the enterprise. It claims that the Undertakers had received "a Grant of the Towne of Boston of Three Thousands Acres of Land and woods and the like from the Town of Lynn . . ." (Suffolk Deeds, XIV, 42.) It is possible that the now lost Lynn town records contained confirmatory data. While later instruments of sale and mortgage often refer to "Iron works land," and not infrequently in combination with "the common lands of Lynn," the citations give no indication of total area, precise location, or original grantor.

6

Braintree Works

WHILE THE SITE of Winthrop's ironworks construction has been
ardently debated for more than a hundred years, it is now clear that
he began work at Braintree. In all likelihood, that town's bog iron
ores figured in Winthrop's plans when first he went to England to
raise the capital for a New England ironworks. Back in America,
he was still persuaded of their fine promise. Whether at the Undertakers' urging, or on his own, to make assurance doubly sure, he
improved the interval in which bad weather ruled out actual building
work by further exploration of possible sites for his ironworks. His
prospecting carried him far, and in it he had the help of the skilled
experts he had recruited in England. Winthrop and his men went
everywhere that finds of iron had been rumored, checking on every
possibility of iron ore. The results of a survey, which must have been
as strenuous as it was far flung, were embodied in a report to the
Company which made it plain that he still preferred Braintree.[1]

The criteria for selecting an ironworks site are explicit. The proposed plant would require a supply of ore and charcoal timber, a
site whose topography was suited to the construction of blast furnace
and forges, and availability of common labor, of workers' housing,
of teams and draft animals. Quality of ore was important, of course.
It was also necessary to consider, however, the general situation of

[1] *Winthrop Papers*, IV, 425–26.

the area in which the ore was found. There was ore, some of it un-
familiar to Winthrop's workmen, but all of it presumably more or less
promising, at "Greensharbour," "Richman Iland," "Pascataway,"
"Agamenticus," "Sako," "Blackpoint," an unnamed location about
thirty miles westward, "Nashaway," "Cohasset," "Woburne," and
"many other places." The type of ore at "Pascataway" and "Agamenticus" is not specified, but, according to Winthrop, specimens had
been sent to England for trial. Of the "Sako" and "Blackpoint" ores
we are told only that they were extensive and of good quality. The
deposits at "Plimoth," "Greensharbour," "Nashaway," "Cohasset,"
and "Woburne," like those at Braintree, were bog ores. All of these
places were remote and little settled. Unskilled labor would be scarce
and expensive. Ore, wood, and land would be free for the asking, but
housing and all the other basic facilities would have to be developed
from scratch.

At Braintree, on the other hand, there was bog iron ore which had
been tested and found good. Similar deposits were available elsewhere,
but here, in the judgment of Winthrop's workmen, was enough to last
twenty years. Here, also, were two good building sites. True, there
were disadvantages. All of the land in the vicinity was already in private occupancy. Boston had promised an allotment of three thousand
acres from its common land, but this was to be more than a mile from
the proposed plant location. Wood, ore, and other necessaries would
have to be purchased or hauled in from some distance. There were
advantages to offset these drawbacks, however. Braintree, Winthrop
assured his sponsors, was "the heart of all the English colonies."
Unskilled help was plentiful and near at hand. Teams and housing
for the workmen could be hired. While Winthrop was asking the
Undertakers to choose between starting work in the wilderness and
beginning at Braintree, he nevertheless concluded his report by indicating that "necessity" seemed to drive the Company to "accept of"
the latter place.

It is not clear that the Undertakers expressed their concurrence
in this judgment. It is all but certain, however, that, during 1644,
Winthrop erected a blast furnace at a site which is today within the
limits of Hall Cemetery, in West Quincy.[2] There, on land which

2 There has been much confusion and much controversy as to where Winthrop began
his furnace construction work. Zealous local historians of Lynn, Braintree, and Quincy

Winthrop purchased, in conjunction with three local investors referred to as "Assistants in the Iron worke," from Edward Hutchinson,[3] his imported craftsmen built a furnace which made iron, and so must qualify as the first successful blast furnace set up within the present limits of the United States.

Work was presumably started as soon as the weather permitted, probably early in the spring of 1644. By mid-May of the following year, Winthrop had completed construction of the furnace and had got in a good stock of ore, charcoal, and wood. The town of Boston had come through with its generous grant-in-aid. In January, 1644, a general town meeting awarded Winthrop and his partners three thousand acres of the common lands of Braintree "next adjoyning and most convenient for their said Iron worke." Dorchester assigned them fifteen hundred more.[4] The General Court was well pleased. Asking private citizens to subscribe the additional capital needed to build the forge, it assured them that the ironworks was very successful "both in the richnes of the ore & the goodness of the iron" and promised to be of great benefit to the whole country.[5] Evidently, at least some iron had already been made.

Back of the attainment of this degree of success, however, had been the surmounting of many difficulties. There had been at least some unfilled workers' billets. Winthrop had brought over at least a

---

have gone to great pains to "prove" the claims of their respective towns to a "first" which really belonged to Virginia. Partisanship, together with the neglect of some, and the misinterpretation of other, documentary data, led to untenable conclusions. One claim for 1643 as the date of actual ironmaking operations at Lynn centers on an iron pot alleged to have been procured as the "first casting" from the new plant. (Alonzo Lewis, *History of Lynn, Essex County, Massachusetts* [Boston, 1844.]) In my judgment it rests on a highly questionable family tradition. Analysis of filings from the pot, now on display in the Lynn Public Library, has proven that it was made at Hammersmith. Its date of manufacture is, however, unknown. The records of purchase of the furnace site and of the furnace building to December, 1644 (*Winthrop Papers*, IV, 424, 498–99; V, 359); the date of the Boston land grant to Winthrop (*Second Report of the Record Commissioners of the City of Boston* [Boston, 1877], 77); and a very early statement that the ironworks was begun at Braintree (Edward Johnson, *Wonder-Working Providence of Sions Savior in New England* [London, 1654] [Reprint, Andover, 1867], 207–208), all seem to me to assign the Massachusetts priority to the Furnace Brook location.

[3] *Winthrop Papers*, V, 359.

[4] *Second Report, Boston Record Commissioners*, 77; *Winthrop Papers*, V, 7. It is possible that the Dorchester grant was subsequently canceled.

[5] *Mass. Records*, II, 103.

"miner," a "founder," a "finer," a smith, and a clerk. He was still short a "bloomer," unless the joining of the workmen's heads together had successfully converted a smith into one, in accordance with Foote's suggestion. This deficiency may have been quite crucial if plans called for the bloomer to make the iron needed in furnace construction right from the ore in a presumably crude and temporary direct-process forge. It seems more likely that Winthrop would have used English wrought iron in the construction of what appears to have been originally conceived as an indirect-process plant. No documentary or archeological evidence for a Winthrop bloomery has been uncovered. The five tons of imported stone doubtless destined for the furnace hearth and stack lining, which Doctor Child had arranged for, had probably arrived soon after the *An Cleeve* reached port. Vastly greater quantities of the native granite of which the furnace mass was built had to be hauled, cut, and laid, however, and this was no easy job. Neither was the getting in and working up of the timber for the charging bridge, the bellows, their frames, and the water wheel which was to drive them. A dam had had to be built. Winthrop and his men must have been busy indeed as they pushed construction, felled timber, made charcoal, and gathered in ore.

The operation was as costly as it was difficult. In the *Winthrop Papers* there is a stray item endorsed in Winthrop's handwriting as "Willm. Osburnes account Clerk of the Iron Worke." It covers receipts and expenditures from an unspecified date, presumably that on which work was begun, to December 7, 1644. By that time £390 had been spent, in addition to certain notes of Winthrop's.[6] When the General Court asked for additional capital investment in the following May, it pointed out that by then between £1200 and £1500 had been expended on the ironworks. If the Court was well informed, it would appear that Winthrop's initial capital of £1000 was already exceeded, and the Undertakers were still a long way from the complete indirect-process plant which doubtless had been envisaged from the very beginning.

Besides getting the furnace going, Winthrop had been busy in persuading local residents to join the English capitalists in the iron-making enterprise. Here his success had been only moderate. In Osborne's account, Winthrop himself was down for £189, plus his

6 *Winthrop Papers*, IV, 498–99.

notes in an unspecified sum, "Mr. Tinge" for £54, Henry Webb for £49, "Mr. Welles of Roxbury" for £17, "Mr. Huise" for £23, and "mager Sedgwick" for £29. "Tinge," "Huise," and Sedgwick were the three "Assistants" of the basic land purchase. The sum of their contributions, added to the £1000 that Winthrop had brought back from England, gives a total that falls about halfway of the range cited as having been expended to date when the General Court made its plea for further support. And these were apparently all the local shareholders there were.[7] The Court's appeal, in other words, seems to have fallen wholly on deaf ears. Governor Winthrop's *Journal* for the year 1645 has a statement that "only some two or three private persons" joined in the ironworks enterprise. This, to judge from the Osborne data, was a low estimate. Even as given by Osborne, however, the number of local investors and the amount of their subscriptions fell far short of the "good proportion, (one halfe at the least)," which the Court had recommended that Massachusetts residents assume.[8]

True, the men who saw fit to invest in the ironworks were men of considerable stature and substance. "Mr. Tinge" was William Tyng, a prosperous Boston merchant, who died leaving an estate the largest on record in the Colony to that date and including an investment in the ironworks which totaled £300 when an inventory was taken in 1653. Henry Webb, another Boston merchant, we have already encountered. He will play a role of some consequence in the story that follows. The "Mr. Welles of Roxbury" was probably the Reverend Thomas Weld or his brother. "Mr. Huise" was doubtless Joshua Hewes, nephew and attorney to Joshua Foote. Robert Sedgwick was the founder of the Ancient and Honorable Artillery Company, and eventually major general of the Colony.[9]

Tyng, Webb, and Hewes carried on their investment in the Company, doubtless adding, as Tyng clearly did, to the sums indicated in the Osborne account. Winthrop's own financial stake, for reasons which will soon be apparent, was not long continued. We have no other data on Sedgwick's participation, but, presumably, neither he nor the Welds advanced additional capital. Thus, there seems to

[7] Suffolk R. & F., 125–2.
[8] *Mass. Records*, II, 103.
[9] Savage, *op. cit.*, IV, 48, 357–58, 456 *et seq.*, II, 407–408.

have been not only a small movement to get aboard the ironworks bandwagon but some tendency to get off it. Even while these men were active, there was, despite the fact that they were all at least moderately well off, a deficit of about £30 as of December, 1644. Despite their prominence, there was clearly no rush on the part of others to join them in the following year. The conclusion that people who were close enough to the scene to know what was going on took a less than rosy view of Winthrop's ironworks is almost inescapable.

While we do not know any of the particulars, Winthrop himself was apparently not too well pleased with his new business venture. Reading between the lines of a few letters and other documents, and generalizing from the records of others who attempted to set up ironworks in still less than well-settled areas, one might gather that the task of managing the ironworks in its infancy, and with less than adequate capital, was more than a little strenuous. The talented promoter must have been disappointed in the rather poor results of his attempts to interest his fellow residents in his project. He certainly ought to have felt some awkwardness in the squeeze position into which the Undertakers' demands for more and more favors and privileges had pushed a man who happened to be both their agent and the son of the Governor of Massachusetts, committed to the Colony's general religious and social principles. Besides all this, and doubtless more or less closely related to it, there were, so far as Winthrop the man was concerned, the tugs of alternative fields of interest and, in the case of the Company, very evident dissatisfaction with his record of performance.

Even during 1644, when he should have had more than enough to occupy himself in the ironworks, Winthrop was already looking to other pastures. First of all, there was the graphite at Sturbridge. His hopes for exploiting this material had been as long cherished, perhaps, as his dreams of a New England ironworks. Robert Child, we know, had been told about it and had expressed a strong interest of his own as early as the summer of 1643. In November of the following year, Winthrop's agents were carefully buying from the Indians the graphite-bearing lands at Tantiusques soon after he had obtained from the General Court a grant of the hill "in which the black lead is."[10]

[10] *Winthrop Papers*, IV, 395, 495–98; *Mass. Records*, II, 82.

There were also ambitions in other nonferrous fields. A copy of an undated petition to the General Court survives in the *Winthrop Papers*. Probably drawn up in this period, it undertook to show that the rocky hills of Massachusetts might contain, besides the "most necessarie mynes of iron (which some with much Cost and Difficultye haue attempted, and in a good measure accomplished to the great benefitt of these plantacions)," lead, tin, copper, and other no less valuable metals. Their exploitation would be slow, painful, and costly. As inducement for taking on the task, Winthrop sought privileges little different from what the Undertakers had asked for and received before taking up the work in iron. Among the concessions sought were rights in mineral-bearing land, freedom from taxation, and the privilege of taking privately owned wood and building materials at prices set by arbitration.

Such encouragement was justified by a citation of the problems involved in metallurgical work so eloquent as to make one wonder if actual experience at Braintree had not sharpened Winthrop's quill pen. Mineral matters, he wrote, took many ingenious heads and hands and full purses. They were "heavy in managing." To be successful in Massachusetts, where skilled specialists had to be imported, where, as in all places, the sale of the product was "verie hasardous," and where "hands require more than ordinarie incouradgment," to say nothing of the difficulty of seeking out ore and transporting the finished metal in an all but impenetrable wilderness, clearly demanded special concessions from government.[11]

Finally, there is at least a possibility that Winthrop, even as he carried on the Undertakers' work, was considering setting up ironworks of his own. In June, 1644, with Braintree Furnace presumably little more than started, he petitioned the General Court for permission to establish an ironworks at Pequot (New London). While the petition mentioned the last court's grants in encouragement of ironworks, it is not clear that Winthrop was filing it in behalf of the Company of Undertakers. Since it had a monopoly, one might assume that this must have been the case. In neither the petition nor in what appears to be the reply it elicited from the Court, is the Company so much as named, however. Indeed, in the reply, if such it is, there is no mention of ironworks. Winthrop was merely given the right

[11] *Winthrop Papers*, IV, 422–23.

to establish a settlement at Pequot with as many as chose to join him, subject only to a restriction that a sufficient number be assembled within three years.[12]

By about the end of 1644, we know, Winthrop was resolved to quit the job of managing the Company's ironworks. Whether from the pressures of the ongoing ironworks problems, from the lure of new fields for his talents, from disagreement with his English principals, or from their disappointment in his accomplishments, he commissioned Emmanuel Downing, his lawyer uncle, to report to the Company that he wanted to be replaced, either by Downing or by another suitable man.[13]

When Downing reached London with the news early in 1645, he learned that the Undertakers had been doing some pondering of their own and had independently reached the conclusion that Winthrop was not the man for the job. They announced that they had resolved to send over Richard Leader as their agent. Of him Downing reported to his nephew, "You know the man. he lived in Ireland, he is a perfect Accountant, hath skill in mynes, and tryall of mettals. . . ." Then, after summarizing the terms on which Leader was to be hired, Downing gave Winthrop the closest thing we have to an explanation for the supplanting by a new agent of the original promoter of the ironworks, and the son of the first magistrate of Massachusetts—

when I perceiued that they were resolued vpon him; and that yt would be noe advantadge to you for me to haue expressed my dislike of theire way herein, but haue putt more Jeolosies into their heads of you; and when they asked me what I thought thereof, I answeared that you had travayled from East to West from North to South sparing noe Costs or paynes for the discouerie of mynes and fitt places for erecting of Ironworks; and how you obteyned 3000 acres of Boson, 1500 of Dorchester, wherein you have deserved well from them, and that there wilbe great neede of your helpe though they send one never soe sufficient for the worke, whereto they replyed that they resolued to satisfie you for the tyme past, and to desire your assistance for tyme to come. . . .[14]

Mr. Leader, perhaps for friendship's sake, perhaps because he

[12] *Mass. Records,* II, 71.
[13] *Winthrop Papers,* V, 7.
[14] *Ibid.,* V, 6–8.

could imagine what life in Massachusetts would be like if he went there under a dark cloud of Winthrop family disapproval, had indicated that he would not take the post if John Winthrop failed to manifest his "free Consent and Contentment." Getting assurance from Downing on this point, he quickly concluded the business. The new agent was duly engaged. The Undertakers prepared to write a letter of thanks, with assurance of satisfaction of his outstanding claims, to the man who not too long before had persuaded them to invest in New England iron and who was now released, to the full or nearly full satisfaction of all concerned, to take up other pursuits.

Obviously, the magnitude of the change demands further attempts at explanation. If our understanding of Winthrop's motives must rest largely on conjecture, we are little better off when it comes to trying to understand what lay back of the Undertakers' change of heart. Presently available physical evidence at the furnace site and stray documentary clues permit at least tentative judgments. Fortunately the data from these two quarters seem to be complementary with respect to Winthrop's selection of the ironworks location. In other areas the picture is less clear.

While Edward Johnson in his *Wonder-Working Providence* wrote that "the Land affording very good iron-stone divers persons of good rank and quality in England, were stirred up by the provident hand of the Lord to venture their estates upon an iron work, which they began at Braintree . . . ," one who visits the recently uncovered base of Winthrop's furnace must wonder if divine intervention in New England ironworks had not run a little thin when it came to choice of the place at which to begin. On a stream known to this day as Furnace Brook, and close to a natural bluff well enough suited for anchoring a charging bridge which extended over the casting area, Winthrop's men erected a blast furnace similar in shape, size, and general construction to that which later was to arise on the bank of Saugus River at Lynn. Granted, modern granite quarrying and concentrated urban building have obviously produced marked changes in the general topography of the area. It seems doubtful, nevertheless, if a pond created by damming Furnace Brook could ever have stored enough water adequately to power the furnace water wheel. Similarly, one cannot today estimate the extent or quality of the bog ores which presumably came from low and swampy areas stretching to the south and west. Doubt must arise as to the experts' claim of enough to last for twenty years, however.

Braintree Furnace was probably in production in the late spring of 1645. Two years later, by which time the furnace at Lynn was in blast, it was reported that "after another Blowing" the Braintree location was to be abandoned for lack of ore.[15] When in 1650, on the occasion of still another change of management, an inventory of the Company's property was taken, the furnace at Braintree was not even mentioned.[16] It had by then been abandoned. It had not been forgotten by the Undertakers, however. Some fifteen months later they were writing their agent, "if the water be nott as yett brought to braintry furnise lett it be done per the first" and "we conceiue if more water be brought to Braintry furnace it shall then not be necessary to take down & removed: but that it was only ordered soe in Respect for want of water: . . ."[17] All of these items have to do with developments after Winthrop's time. Vague as they are, they suggest that the furnace can hardly have been properly located in the first place. One may infer that a poor production record, particularly in time of drought, may have had much to do with the Undertakers' willingness to see Winthrop replaced.

In all likelihood, too, they must have been disappointed in their agent's negotiations with the General Court for grants and immunities. For this we have no documentary evidence. We do know, however, that the English investors were very much troubled when first they learned that the land on which the ironworks was to be built had to be purchased. They grudgingly gave their consent. They made it plain, however, that land purchase was to be tightly restricted to the two or three acres needed for the work site proper.[18]

So far as we can tell, Winthrop's sponsors had little with which to

---

[15] *Ibid.*, V, 140; *Mass. Records*, II, 103.

[16] *Essex R. & F.*, I, 294–95.

[17] Ironworks Papers, 29, 39. (The typescript reads "remaned" for the "removed" of the original manuscript.)

[18] *Winthrop Papers*, V, 6. This decision of the Undertakers may have upset a few plans for profiting in real estate transactions. Downing, reporting it to Winthrop, added the somewhat wistful comment, "I pray therefore keepe mr. Hutchinsons land for yourselfe or me, which I suppose wee may improve to good advantadge." William Tyng may have had a related reason for being interested in the ironworks. In October, 1641, he had bought one hundred and sixty acres of Hutchinson land that must have been close to Winthrop's furnace site. *(Suffolk Deeds, I, 23.)*

find fault in the overall costs of his ironworks activities. If an expenditure of some £1300 had seen workers and materials transported to New England, the former maintained through a winter of inactivity, extensive site explorations carried out, and a furnace built, stocked, and going, Winthrop had not been extravagant. The detailed accounts of the State's ironworks in the Forest of Dean under the Commonwealth give £780 as the cost of a large blast furnace. This one may have been built on the site of a furnace demolished in 1650, with resultant savings on construction costs, and was built without charge for timber, since the Forest belonged to the State. Making crude calculations to cover these items, a £1000-figure seems fair for the cost of furnace building. Since the high cost of labor in Massachusetts probably canceled out the lower costs of certain building materials, if Winthrop got his furnace set up for £1000, he would have been doing well.

Winthrop's performance as a business manager is harder to estimate. The forge was not built by May of 1645. Until a finery and chafery plant was available to convert the cast iron from the furnace into wrought, the only ironworks products that could have been sold would have been cast domestic articles such as pots, skillets, andirons, firebacks, and the like. Such items were later produced at the Company's works at Lynn. They there accounted for only a small fraction of total sales. Until Winthrop's furnace got its complementary forge and he could sell bar iron, it may be taken for granted that expenses must have exceeded income. Winthrop can hardly be held responsible for this state of affairs.

In the eyes of the English capitalists, however, things must have looked grim. They saw a continuing drain of operating capital coupled with the need for more fixed capital with which to get the ironworks complete. They must have been aware of the obvious reluctance of Massachusetts residents to join in their business venture. If all this were combined with a badly located furnace, the general ironworks situation must have appeared so uncertain as to make the Undertakers wonder if adding to their investment might be a matter of pouring good money after bad. We know Winthrop ran short of funds and fell back on bills of exchange drawn on the Company. Downing had apparently taken several to London. In reporting to Winthrop the Company's plans for a new manager, he wrote that he had presented but one of the bills, one for £1000. The Undertakers, obviously doubting if so much had been expended, announced that they would pay but £400, and that any balance was to be made available to Leader. If Winthrop had in fact paid out more than this sum, the balance would be paid him with interest.

The tightening of the English group's purse strings, and their decision to send over a new agent, suggest that they regarded their New England investment as sufficiently precarious to require drastic steps. From the Undertakers' point of view, Winthrop had failed to deliver on the project he had doubtless presented to them in glowing terms when he was seeking their financial support. If we must tread warily in pronouncing Winthrop a poor engineer, we may fairly safely ascribe to him a failing common to promoters—that of underestimating costs and overestimating both the amount and the speed of return on investment.

If the reasons for Winthrop's departure are not wholly clear to us, they can but have been compelling. Only a crisis would drive a group of worldly capitalists to a step which might have had disastrous consequences for their enterprise. Winthrop was well connected and enormously popular. To the people of Massachusetts he must have been "Mr. New England Ironworks." If he had not been able to draw out much local financial support for the ironworks, a successor who was a stranger could expect to fare less well, and the Undertakers must have known it. One must wonder if the Company would have chosen to replace Winthrop without his consent. As things worked out, it was doubtless quite fortunate that both sides were agreed that the ironworks needed a new manager, and the transition was easily effected. The letter of thanks to Winthrop which was mentioned above has not survived. Thanks were conveyed, however, in the formal instrument by which Winthrop was asked to hand over works, stocks, and accounts to Leader.

The voluminous Winthrop Papers are strangely silent on the details of the financial settlement between the Undertakers and their first agent. Downing had informed Winthrop that Leader was engaged at an annual salary of £100, and that had he been a smarter bargainer, he might have had £150. He therefore suggested that Winthrop ask for "noe lesse than 150li per annum for these three years."[19] This

19 *Winthrop Papers,* V, 7, 11, 27.

would presumably have covered some of Winthrop's promotional activities as well as his actual work on his return to New England. Some such sum must have been paid him, and in cash, since neither then nor thereafter does he figure as a shareholder in the Company. His original working arrangement with the English investors is wholly obscure. From what we infer transpired as he was leaving their service, one must suspect that they were rather loose. Evidently, he was now reimbursed in full for all he had spent on the ironworks, and given a sum agreed upon between him and the Company as fair compensation for his services. Whatever his formal connection with the Undertakers had been, it was now wholly severed.

There is not the slightest evidence of ill feeling in either quarter. Relatives and friends subsequently wrote Winthrop from time to time about developments at the ironworks. It is logical to assume that he never lost interest in the progress of the enterprise. At one point he helped out in the Company's relations with its workmen, or in the magistrates' relations with the Company's workmen, and received an appropriate letter of thanks from the Undertakers. Leader became his friend; and Winthrop's intimate friend, Child, worked closely with the new agent until his embroilment in political affairs brought his metallurgical work to an untimely halt. So far as we can tell, Winthrop simply withdrew from the Company he had been instrumental in founding and from the management of the ironworks he had started, and happily took up other activities.

If there was no ill feeling, so there was no stigma of failure, no blemish on his reputation as scientist, engineer, promoter, and business man. Given Tenhills, his father's Mystic River farm, in 1643, along with twelve hundred acres on Concord River, Winthrop might have settled down to the life of a country squire. Instead, private concerns of extraordinary multiplicity led him to work and live both in Massachusetts and Connecticut, and an impressive public career was to be crowned with years of service as the chief magistrate of the latter colony. Separation from the ironworks emphatically did not leave him unemployed!

Long before he turned to ironworks, he had become interested in the Pequot country, which had been conquered by the joint efforts of Massachusetts and Connecticut in 1637, and was claimed by both. In 1640 he obtained the grant of Fisher's Island, at the entrance of

Long Island Sound. Under the 1644 authorization from the General
Court mentioned earlier, he began a plantation at Pequot, the present
New London, in 1646. Despite this, he was, in 1648, launching plans
for making salt in Massachusetts.[20] Until his father's death in 1649
he divided his time and energy between his old and new areas of
interest. That year he decided to settle permanently at New London,
and since, in the interim, the New England Confederation had
awarded the Pequot region to Connecticut, it was ncessary for him
to resign his Massachusetts magistracy and tie his fate to the colony
to the south. New London became his home, although his Massachusetts properties and loyalties took him to Boston frequently, and
offers to settle in New Haven, Providence, and even New Nether-
lands, on highly advantageous terms, were often tempting.

His public career in Connecticut was distinguished. A magistrate
there even before he had finally determined to live at New London,
he was elected assistant in 1651 and governor in 1657. His popularity
and ability caused Connecticut to change its rule on succession in the
chief magistracy after he had stepped down for one term in 1658 as
deputy governor. From 1659 to his death in 1676 he continued in
the governor's seat without interruption. His political accomplishments were many, not the least among them his procuring a charter
for Connecticut in 1661 which assured the continuance of its independent government in an enormous area extending all the way to
the Pacific Ocean.

Owing partly to the pressures of his public responsibilities and
partly, in all probability, to the continuing impulse to spread himself
too thin, Winthrop's private business concerns were less than richly
successful. In the metals field, for example, an area which intrigued
him all his life, and which is of special concern to us here, his record
of achievement was spotty. The graphite at Sturbridge, for which
Doctor Child had given assurance of an ample market, and specimens
of which Leader had pronounced to contain silver, got exploited by
Winthrop in conjunction with a number of partners. Despite long
and costly efforts, the venture was a failure. The mine's wilderness
situation made it difficult to get and keep workmen, and transportation costs of such product as was forthcoming were excessive.[21]

[20] *Mass. Records*, II, 229.
[21] George Haynes, "The Tale of Tantiusques," American Antiquarian Society, *Proceedings, New Series*, XIV, 471–97.

Granted rights by Connecticut in any mines of lead, copper, tin, antimony, vitriol, and the like which he might discover, Winthrop always cherished the hope of striking it rich in the minerals of the Pequot country, but things worked out otherwise. In 1655 he made another attempt in ironworks, this time at the outlet of Lake Saltonstall, near today's main highway between New Haven and Branford. The story of the New Haven Ironworks is told in another chapter. It is enough to indicate here that despite good encouragement from public authorities and the help of substantial partners, this new project also turned out to be a faltering venture, from which Winthrop removed himself in 1657.

In other areas close to his heart, Winthrop came to acquire a fine reputation. In the practice of medicine his services were in great demand and freely given. The Winthrop Papers abound in letters in which his medical opinion and prescriptions were sought by rich and poor in Massachusetts, Connecticut, and even New York. His interest in science, particularly in chemistry and alchemy, remained strong throughout his life. His correspondence with many of the world's great scientists was as extensive as long continued. During his stay in England as agent of his Colony, he was elected to membership in the recently organized Royal Society. To it, as we have seen, he made interesting, if considerably less than earth-shaking, contributions. During the years he spent in Connecticut, he continued to send to England letters on all kinds of natural phenomena, now claiming to have discovered a fifth satellite of Jupiter, now writing of a new method of determining longitude at sea, now of the tides in the Bay of Fundy, and so on, almost without limit. By the time he died, on a visit to Boston, he had accomplished more than enough to merit the designation, "America's First Scientist."

Such was the man who led English America one step closer to the exploitation of iron, the basic metal of industry. Such was the man who sparked the formation of the Company of Undertakers of the Iron Works, and parted company with it before its American venture was more than a puny fledgling. Despite the lack of full fruition of many of his projects, he must be accounted an industrial pioneer of no mean consequence. Great by any standards, and singularly attractive when gauged by those of modern America, he well deserved Morison's accolade: "In John Winthrop were combined as in few

Americans of any age the power to think and the power to do, a creative genius and a quality of leadership, the zest of a scientist and the faith of a puritan, a pioneer's energy and the tenderness of a saint."[22] Whatever his difficulties at Braintree, whatever the cause of his falling out with the Undertakers, it is doubtful that New England could have had ironworks so early without him.

It is difficult to tell to what stage Winthrop had carried things at Braintree by the time Leader took over in midsummer of 1645. It is extremely doubtful, however, if much or, indeed, if anything had been done on the forge. In its May appeal for funds the General Court announced that some £1500 would be needed to "finish" the forge.[23] Now a forge was no cheaper to build than a furnace. At least, it cost the Commonwealth £800 to set up two fineries, a chafery, workmen's housing, a warehouse and a charcoal storage house in the Forest of Dean, where wood was free.[24] Similarly, a forge in Derbyshire is indicated as having cost the same amount in 1644.[25] Unless highly peculiar conditions prevailed in Massachusetts, precious little could have been accomplished on a forge needing £1500 to be "finished." Since a change in management was impending, Winthrop would have had little reason to press new construction, and reason the less if his furnace building job had been less than successful. The forge at Braintree was erected on the Monatiquot River at some considerable distance from the furnace. Given the circumstances just outlined and the fact that we have no record of land purchased by Winthrop in the Monatiquot River vicinity, it looks as though Braintree Forge had been built by Leader rather than by Winthrop.

Braintree Furnace, as we have seen, probably operated not much beyond 1647. Despite the Undertakers' hopes that improvement in the water supply would bring it into production again, there is no evidence that it did. When in 1653 the Company's financial difficulties brought it to grief, there is reason to believe that its third agent, John Gifford, had everything in readiness for a blast at Winthrop's old furnace.[26] The confiscation of the Undertakers' assets ruled it out.

22 Morison, *op. cit.*, 288.
23 *Mass Records,* II, 104.
24 S. P. 18/130/102.
25 Sir George R. Sitwell, "A Picture of the Iron Trade (in the Seventeenth Century)," Derbyshire Archaeological and Natural History Society, *Journal,* X (1888), 35.
26 Ironworks Papers, 193.

There is no evidence that the Company's creditors who took over the property ever got this furnace going. Its production during the indicated relatively brief operating span is hard to estimate. Surviving slag evidence suggests that it may have been respectable.[27] Essentially, however, the furnace must be considered one of the less significant elements in the Undertakers' New England industrial plant.

Braintree Forge was a different proposition. While Winthrop's furnace was working, it doubtless converted some of its product into the bar iron which the Colony so much needed. When Braintree Furnace was abandoned, it joined the forge at Lynn in reducing to wrought iron the sows and pigs that came from the blast furnace which had been erected at Hammersmith in the meanwhile. It became, in other words, a subsidiary branch in the Company's ironmaking operations. Its story is thus best postponed until we consider the Undertakers' main works on the banks of Saugus River.

[27] Rather extensive evidence of clay-mold fragments leads to the same conclusion.

Photograph by Harrison Kerr.

Hammersmith slag pile on the Saugus River.

Photograph by Harrison Kerr.

Ironmaster's house.

By permission of the Massachusetts Historical Society.

John Winthrop the Younger, portrait by an unknown artist, from The Laboratory, *Vol. 24, No. 5 (Fisher Scientific Co., Pittsburgh).*

Photograph by Richard Merrill; permission of the Saugus Ironworks Restoration.

Remains of the original blast furnace at Hammersmith. Sills of bellows frame and part of wheel pit. Tree-covered slag dump in background.

Photograph by Richard Merrill; permission of the Saugus Ironworks Restoration.

Remnants of original water wheel, Hammersmith blast furnace.

Photograph of a painting in the restoration museum; courtesy Saugus Ironworks
Restoration.

Hammersmith in 1650, as conceived by restoration architects.

Photograph by Richard Merrill; permission of the Saugus Ironworks Restoration.

The making of charcoal. Painting by Charles H. Overly.

Photograph by Richard Merrill; permission of the Saugus Ironworks Restoration.

Ironmaster's house and restoration museum.

Photograph by Richard Merrill; permission of the Saugus Ironworks Restoration.

Ironmaster's house, interior.

Photograph by Richard Merrill; permission of the Saugus Ironworks Restoration.

The forge (at left) and rolling and slitting mill.

The blast furnace, with forge and rolling and slitting mill at right.

Photograph by Richard Merrill; permission of the Saugus Ironworks Restoration.

The "rock mine," as it appears today at Nahant, Massachusetts. This igneous rock was used as fluxing material in the Hammersmith furnace.

Photograph by Richard Merrill; permission of the Saugus Ironworks Restoration.

The chafery, forge interior.

Photograph by Richard Merrill; permission of the Saugus Ironworks Restoration.

(From left to right) Ironmaster's house, restoration museum, warehouse, blast furnace, forge, rolling and slitting mill. Saugus River in foreground.

Photograph by Richard Merrill; permission of the Saugus Ironworks Restoration.

The hammer, forge interior.

Photograph by Richard Merrill; permission of the Saugus Ironworks Restoration.

The bellows, for the blast furnace.

Photograph by Richard Merrill; permission of the Saugus Ironworks Restoration.

Rolling and slitting mill, interior. Rollers and cutting disks in foreground, reverberatory furnace in background.

Photograph by Richard Merrill; permission of the Saugus Ironworks Restoration.

Forge interior; the fineries.

Rolling and slitting mill, interior. Rolling and slitting frames in fore-
ground, cog and lantern wheels in background.

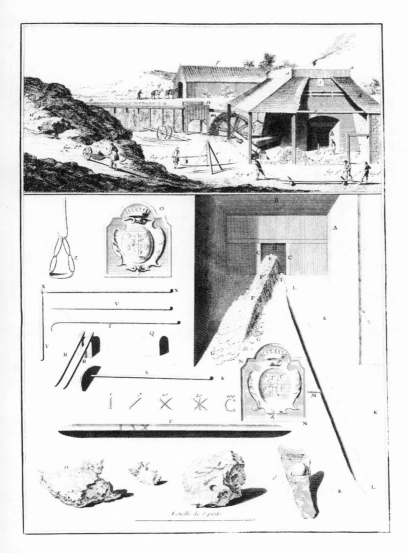

Eighteenth-century blast furnace. From Courtivron et Bouchu, "Art des Forges et Fourneaux a Fer," Descriptions des Arts et Metiers.

Operations in the rolling and slitting mill. From Swedenborg's De Ferro *(1734) in Courtivron et Bouchu, "Art des Forges et Fourneaux a Fer,"* Descriptions des Arts et Metiers.

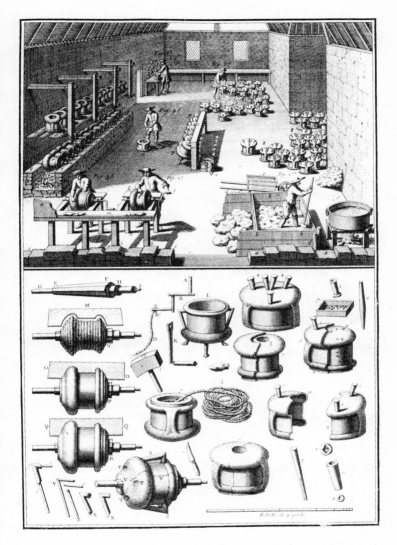

The making of molds for iron pots. From Courtivron et Bouchu, "Art des Forges et Fourneaux a Fer," Descriptions des Arts et Metiers.

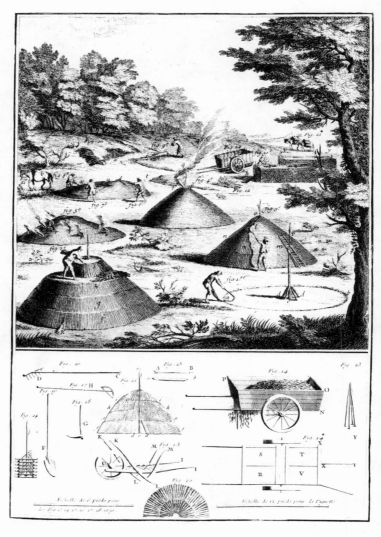

The making of charcoal in 1760. From Duhamel du Monceau, "Art du Charbonnier," Descriptions des Arts et Metiers.

PI. II.

Alchemical apparatus of the type presumably used by Winthrop and his scientific colleagues, from John Andreas Cramer, The Art of Assaying Metals . . . *(London, 1741).*

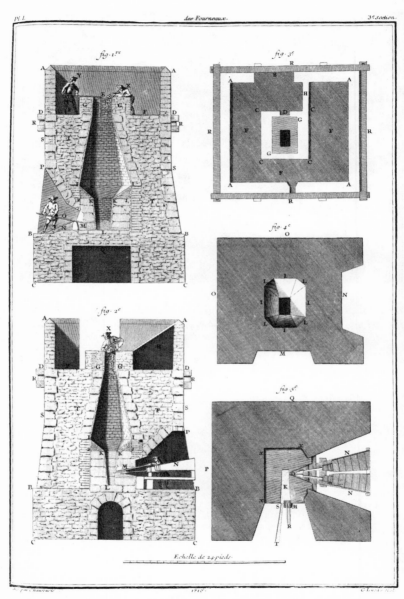

Early eighteenth-century French blast furnace. From Courtivron et Bouchu, "Art des Forges et Fourneaux a Fer," Descriptions des Arts et Metiers *(1762).*

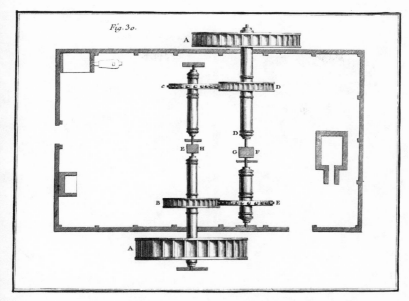

Plan of a rolling and slitting mill. From Swedenborg's De Ferro
*(1734) in Courtivron et Bouchu, "Art des Forges et Fourneaux a
Fer,"* Descriptions des Arts et Metiers *(1762)*.

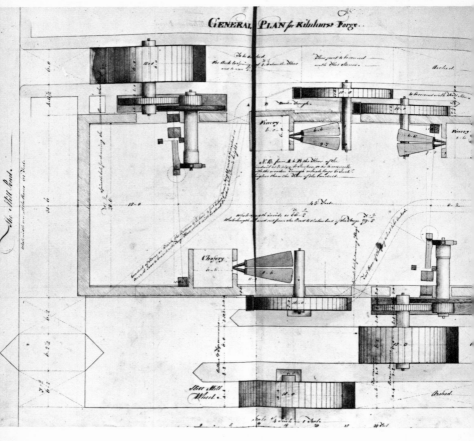

By permission of the Royal Society.

John Smeaton's general plan for Kilnhurst Forge. The Kilnhurst plant was more elaborate, but otherwise had much in common with the Saugus works.

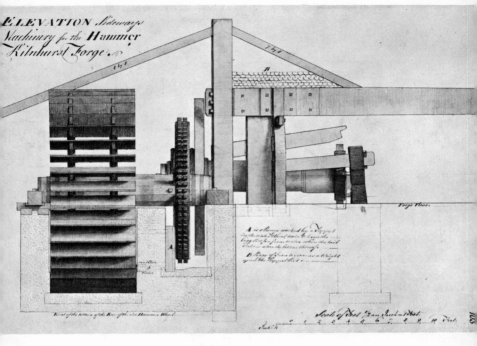

By permission of the Royal Society.

John Smeaton's elevation of the hammer machinery for Kilnhurst Forge.

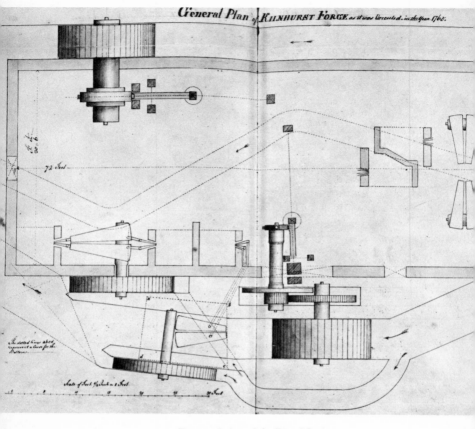

By permission of the Royal Society.

John Smeaton's general plan of Kilnhurst Forge as executed in 1765.

Photograph from the author's collection.

Remnants of Raynham Forge, pioneer ironworks in the Plymouth Colony, as seen in the early 1900's.

Photograph from the author's collection.

Raynham Forge, pioneer ironworks in the Plymouth Colony, as seen in the early 1900's.

7

A New Agent
and a New Site, Lynn

RICHARD LEADER'S QUALIFICATIONS as a "perfect Accountant" doubtless derived from a background of experience as a small merchant
working mainly in the Irish trade. In all likelihood the son of David
Leader of Speldherst, Kent,[1] we first encounter him shipping foodstuffs, tallow, stockings, and pike and musket staves from Limerick
to London. By then he was describing himself as "of Salehurst, co.
Sussex, gent aged 28."[2] For some time prior to 1641 he was factor
and agent at Limerick to a merchant named Beale. Then, like many
workers in the vineyard of economic imperialism at this and later
junctures, he fell victim to political disturbance. A ship laden with
malt consigned to Leader reached Limerick almost on the eve of the
Irish uprising. He had barely got the cargo off and replaced with hides
and other merchandise bound for France when, to use his own words,
"the Toune rose up and was in rebellion and the Castle of the then
King was by them afterwards seized, the goods of this deponent's by

---

[1] Prerogative Court of Canterbury—Wills, Aylett, 257; Wills, Laud, 155. There was
another Richard Leader of Salehurst, who died in 1614. The existence of two contemporaneous Richard Leaders may occasion doubt as to the soundness of the identification
of Richard the "agent" in the English data summarized in the text. In most of the
documents cited, the agent's flowery, elegant, and legible signature leaves no question
as to authorship. Furthermore, the wills referring to another Richard Leader make
no mention of George Leader, a brother of the ironmaster, who accompanied him to
Ireland and eventually to America.

[2] H. C. A. 13/53, fol. 118.

the Rebelles seized and this deponent . . . was forced to flee with his Wife and familie."[3] Obviously, it was time to go home.

Back in England, Leader was far from abandoning all hope of success in Irish trading ventures. In 1643, calling himself "Richard Leader of London merchant," he put twenty-five marks into Irish land, and, some years later, bought still more in conjunction with David Leader. In 1644, however, he was engaged in the coal trade between Leith and London, in partnership with one Robert Petley. He doubtless endeavored to carry on other mercantile activities. While we have no way of knowing how lucrative such home-country concerns may have been, it is not at all unlikely that when the Under-takers turned to Leader, they were dealing with a man whose "normal" business had been interrupted by the Rebellion, who had tasted the fruits of carrying on business in one "backward area," and who doubtless was more than eager to try his luck in another.

Where Leader obtained the "skill in mynes and tryall of metalls" referred to in Downing's letter to Winthrop is a mystery. It is possible that he had had some exposure to the various ventures in which English capitalists endeavored to develop Ireland's mineral resources during the 1630's and 1640's. Certain of these in the ferrous field have been mentioned earlier. There were some in other areas. In 1631, for example, Thomas Whitmore and William Webb secured a patent which gave them all the mineral works and mines of copper, gold, silver, and the like in Munster for twenty-one years. By 1633 they had spent £4000 in importing English workers and operating mines in Tipperary, and all to no profit.[4] Leader may have been employed on this or some similar project as a young man. It is doubtful, however, if he would have had much time after 1637 to spare from his trading duties for mining operations. It is conceivable, too, that in his youth he had had contact with ironmaking in Sussex.

Somewhere along the line, we know, Leader had become something of a scientist. This was probably the initial basis of his friendship with young Winthrop and Doctor Child. We are told that he had a good library, one which Child pronounced more "curious" than his own, especially in the religious field. There was, however, quite a gulf between the work of a "library" or "laboratory" scientist and that

[3] H. C. A. 13/63, August 8, 1650.
[4] C. S. P. Ireland, 1663–65, pp. 4–5, 67.

of a practical metallurgical engineer. The Undertakers may have thought that they were getting an experienced metals man. In some respects, however, they were getting, well, another Winthrop. This man's skill in the "tryall of metalls" was not sufficient to keep him from announcing, after an analysis of specimens from Winthrop's graphite deposits, that the latter's "leade oare" was a "silver Myne."[5] His practical ironmaking experience, even after some years of work in Massachusetts, was not adequate to spare him the necessity, on one occasion, of giving what appears to modern eyes as almost the epitome of a theoretician's response to a basic technological question. In 1655, Winthrop asked Leader for working dimensions and other data for the furnace he was then about to build at New Haven. His successor in the management of the Undertakers' ironworks replied that he was at a loss, since the book containing the data was packed away in a trunk and unavailable! As an alternative, pending his finding of the book, he suggested that Winthrop turn to Osborne, who presumably still had the letter Leader wrote him in 1650, and according to which he had built the hearth "that made both the most Iron and best yield that ever was yet made in New-England."[6]

The resemblance of Leader to Winthrop does not cease with this real but largely book-oriented interest in science. Leader, too, felt the tug of a thousand interests, interests which were rivaling his calling to the ironworks almost as soon as he reached Massachusetts. Like Winthrop, he was rather more liberal and rather more modern in tastes and judgment than many in the Puritan community in which he had now to live. He did not sign the Remonstrance that got Child into trouble. There is little question, however, as to where his sympathies lay, and it was probably more his concern for the Company's interests than his own good sense that kept him from subscribing to an action which could only have amounted to political indiscretion or worse. Certainly, he was less cautious once he had resigned as the Undertakers' agent and, as we shall see, paid dearly for expressing independence of judgment in a society in which conformity was the legally enforced norm.

Whatever else Leader was, he was clearly an "enterpriser," product of the new age of capitalism, willing to break with the old and risk

[5] *Winthrop Papers,* V, 6–7, 21.
[6] *2 M. H. S. Proc.,* III, 192, (Winthrop Papers, 14.125.)

the new. His Irish experiences had probably provided useful precedents for, and perhaps added to the lure of, his new field of operations
in Massachusetts. Similarly, if there were deficiencies in his command
of matters metallurgical, there was, in compensation, the personality,
the bent of temperament, of the merchant "on the make," to use the
slang term. Winthrop's social position was loftier. His family ties and
the responsibilities of his high station must always have tempered his
independence of judgment, though obviously not his willingness to
take risks. Basically, however, Winthrop and Leader were simultaneously scientists and entrepreneurs. They had a common interest in
business and in technology. They were alike under the spell of minerals and of the other natural resources of the New World. Winthrop's
birth may have given him a preferred place on the ladder of capitalistic ambition, but both men were cut from the same cloth.

It is clear, however, that Leader's position with the Company of
Undertakers was different from Winthrop's. The latter had been the
promoter of the whole venture, had been well connected in the old
country and in the Colony, had been part owner as well as agent,
quite on a par with the English investors. Leader, by contrast, was
a salaried manager. It is likely that either in England or upon his
arrival in Massachusetts, he invested some money in the Company.
His status, however, continued to be that of the hired employee subject to the Undertakers' orders. They had agreed to pay him £100
per year, starting March 25th, 1645, to pay for his transportation
and that of his wife, two children, and three servants, to build him
a house and allow him ground for the use of his horses and a few cows.
In return, Leader was to take charge of the Company's New England
affairs for seven years. In all likelihood, these quite generous arrangements were concluded by formal contract.

Despite the date at which his salary started, Leader seems still to
have been in England in May. Indeed, the letter authorizing Winthrop to turn over the Company's assets is dated June 4th and begins,
"According to what we haue formerly written vnto yow we now send
over our Agent, Mr. Richard Leader. . . ."[7] His first local business
transaction of which we have record is dated September 29, 1645.[8]
We must assume that the Undertakers' new agent reached these

7 *Winthrop Papers*, V, 6–7, 21, 27.
8 *Suffolk Deeds*, I, 62.

shores in midsummer of that year. With him he brought a stock of English commodities worth £670. With their proceeds he was expected to carry on the works without further advances from the Undertakers.[9] Unless his predecessor had accomplished more than we have suggested, Leader was taking on a mighty task.

One must wonder what thoughts filled his mind as he made his first inspection of Winthrop's achievements to date at Braintree. Presumably, he saw the blast furnace, finished, stocked, and even perhaps going—but badly situated. With it went a small group of workers, probably quartered in the homes of neighboring farmers, the vast land grants still largely wilderness, and little more. One wonders, too, what transpired when Winthrop told him, doubtless in terms more candid than had ever been used with the English investors, the assets and liabilities of the situation in which Leader now found himself. Finally, one would give much to know what Leader thought, first as he was welcomed by the Governor and other local figures as the representative of their "friends" in England, then as he carried on, with much pulling and hauling, the negotiations we have already summarized. Surviving records offer, alas, no direct and but highly limited circumstantial data. It is hard enough to tell what Leader did, impossible to tell what he thought. No one can doubt, however, that he found the ironworks as "heavy" to manage as Winthrop had.

Since even a properly situated furnace would have been but a poor thing minus its complementary forge facilities, Leader's first job was to build or "finish" the forge. This he did at a site on the Monatiquot River within the limits of Braintree, then as now. The precise location has not been established, but the general area in which he worked is known. Its good and bad features are apparent despite modern building, which, as in the case of Braintree Furnace, complicates evaluation from currently available ground evidence. The forge was situated some two miles from the furnace, and the intervening terrain seems not too well suited for the carting of sows from the furnace. It was by no means uncommon, however, for crude iron to be carried over far longer distances to English forges. There, as elsewhere, the location of fineries and chaferies was mainly determined by availability of water power. A forge on the Monatiquot, a lively little stream even today, could count on a good and constant supply. The river was

[9] Ironworks Papers, 33.

probably also navigable by shallow draught vessels, so that some of the finished products may have been freighted by water. Further, a main highway running from Boston to the south was but a short distance away.

Ideally, of course, the close proximity of furnace and forge would have been strongly desirable. Leader, however, was not wholly free to choose. He had inherited an already standing furnace. If it had run short of water, to build a forge on Furnace Brook would have been insane. To pick up altogether and move to a wholly new location was out of the question. Boston and Dorchester had been most generous in their land grants. The privileges awarded by the General Court all hinged on the construction of complete indirect-process ironworks, furnace *and* forge. With a furnace already set up, well or otherwise, this was no time to abandon Braintree as an Undertakers' ironworks site. Leader, like any man in his situation, can only have decided to shop around and find the best forge site available in the general area in which Winthrop had made his beginnings. That on the Monatiquot River had much to recommend it.

We know next to nothing of construction activities: it is not even clear when Braintree Forge was completed and put to work. It was at the end of September that Leader bought the twenty acres of land along the Monatiquot on which the forge was presumably built.[10] In December, John Winthrop, in a Latin diary of a trip to Connecticut and back, wrote that he "passed over monotaquid at twilight. came by the direction of the noise of the falls to the forge."[11] This suggests that the forge was by then in existence. Would that he had seen fit to indicate to what stage it had been carried! A complete working forge might have been set up in two months' time, but since Leader had had to dam the river, construct three water wheels, set up a finery and a chafery and their bellows, and, last but not least, erect a heavy power-driven hammer set in a massive frame, it may be doubted if all was finished and in operation at the time Winthrop made his brief visit. In all likelihood, however, the forge unit was ready for business in the spring of 1646.

While we have no production data on Braintree Forge and no index as to the quality of its product during the whole of Leader's tenure,

[10] *Suffolk Deeds*, I, 62.
[11] *Winthrop Papers*, V, 54.

we must assume that it operated successfully as long as the furnace was providing its supply of crude iron. When he quit in 1650 the forge consisted of the units mentioned above, and was moderately well equipped with working tools.[12] Unlike the furnace, it was never abandoned by the Company.

If Braintree could not be abandoned as an ironworks site, there was no reason why it could not be complemented by others. Urged on by the shortcomings of Braintree Furnace, Leader apparently was seeking out a likelier spot for ironworks building even as he rushed construction on Braintree Forge. A furnace without a forge was of little value; a forge without a dependable supply of crude iron to be worked up into the bar iron which was the ironworks' chief sales item was at best no more than a precarious industrial unit. Clearly, it behooved Leader to extricate himself and the Undertakers from the awkward position he had seemingly inherited from Winthrop.

We have no record of Leader's search for a new and better site. He must have engaged in much the same kind of location surveying that Winthrop and his men had carried out, tracking down reports of ore deposits, checking on availability of water power, pondering the relative merits of wilderness and settled regions, and keeping an eye open for prospects of sales and transportation of finished products. Ten miles north of Boston, on the banks of Saugus River, in that section of old Lynn which is now Saugus, he found a spot which had been overlooked in Winthrop's survey but which clearly had distinct advantages.

In the general vicinity were low-lying meadows and swamps containing bog iron ore of good quality. Hard by the bridge which carried the main road between Boston and Salem over Saugus River was a kind of natural amphitheatre, so situated that on fairly level land washed by the stream a furnace could be erected with adequate water power and easy charging from a natural elevation rising above the riparian plain. At high tide Saugus River was navigable right up to the site in question. Handy as the place was to the growing towns of Salem and Lynn, Charlestown and Boston, it was not far from the common lands of Lynn, much of them covered with stands of virgin timber promising an almost inexhaustible store of wood for charcoal and construction work. One can easily conjure up a picture of Richard

12 *Essex R. & F.*, I, 295.

Leader standing at the top of the amphitheatre and announcing, "This is it!"

Most of the land in the vicinity was in the possession of one of the most fascinating figures in early Massachusetts history, Thomas Dexter. As early, perhaps, as 1630, Dexter had settled as a farmer on the west bank of Saugus River. He acquired title to some thirteen hundred acres of land, set up a weir in the river, and, it is claimed, built a gristmill as well. Improving these properties should have kept him more than sufficiently occupied; the records indicate otherwise. He had been bold enough to buy all of Nahant from an Indian chief known to the English as "Black will" or "duke William"—a transaction whose validity was hardly universally recognized, as at least three later lawsuits made abundantly plain.[13] He was irascible enough to have been before the courts on numerous occasions. In 1631 he claimed to have been the victim of assault and battery by Captain Endicott, and the General Court so found. Normally, however, Dexter was on the other side, the guilty party. In the summer of 1632 he was fined for insolence to Simon Bradstreet, required to confess his fault, and bound to good behavior. In March of 1633 he was put in the bilbows, disfranchised, and fined £40 for reproach of the General Court, and sedition, although £30 of his fine got remitted in 1638. In the interim, in October of 1633, he had been fined for drunkenness.[14] His postures and claims in many of these proceedings abundantly testify to the fact that this was indeed what Yankees call a "character."

Dexter's business transactions, especially in land, were as complicated as his personal life. One who tries to unscramble them as they appear in court records may well wonder if the myriad clouds in the title to his property ought not to have outweighed whatever suitability certain portions thereof may have had for the building of ironworks. There is no need here to spell out the intricacies. Suffice it to say that by the time Leader appeared on the Saugus River scene, Dexter's property had been mortgaged, in whole or part, to no less than four people, some of it simultaneously, and, since it was mortgaged to both

[13] Alonzo Lewis and James R. Newhall, *History of Lynn, Essex County, Massachusetts* (Boston, 1865), 119; *Essex R. & F.*, II, 43; VII, 124–29.
[14] *Mass. Records*, I, 97, 103, 109, 243.

Massachusetts parties and English residents, more conveniently than honestly. One man's claims had been settled by foreclosure, and the awarding to his assignees of seven hundred acres agreed upon by neutral umpires. Those of the remaining three were still outstanding, including that of Simon Bradstreet, who had loaned Dexter £90 in October of 1639, their clash of seven years earlier forgotten or overridden by business considerations. Despite all this, Dexter held some land which, rightly or wrongly, he took to be his to dispose of.

Of necessity, the purchase of land for Leader's ironworks at Lynn was a highly involved matter. In December, 1645, Leader bought out Bradstreet's claims by paying him £65, the balance still due from Dexter. In January, 1646, he bought, directly from Dexter and for £40, "All that parcell of land neere adjacent to the Grantor's house wch shall necessarily be overflowed by reason of a pond of water there intended to be stopped vnto the height agreed on betwixt them, and also convenient land & sufficient for a water course intended to be erected together wth the land lyeing betweene the ould water course & the new one. As also fyve Acres & halfe in the Cornfield next the Grantors house, & twoe convenient Cart wayes one on the one side of the bargained premisses & another on the other side thereof. . . ." It was presumably on part of the land here being conveyed that Leader set up his ironworks. In December of the same year he bought up the claims of the two remaining English mortgage holders covering Dexter's "farm" and weir, respectively, for £230. Finally, in May, 1647, in consideration of £336 sterling, Dexter sold Leader his entire holdings, including his house, six-hundred-acre farm, outbuildings, the fish weir, and all rights and privileges. Dexter agreed to hand over all the deeds and instruments covering the whole property. The present deed was unusually detailed and its handling, for obvious reasons, more than ordinarily careful. Sealed and delivered before three witnesses, extracts from it were entered at both Essex and Suffolk courts. Further, Dexter got prevailed upon to go before Governor Winthrop and acknowledge in his presence that the whole had been his true act and deed.[15] After months of effort and the expenditure of £670, Leader had finally secured the six-hundred-odd acres

---

[15] *Suffolk Deeds*, I, 14, 15, 29, 52, 69, 70, 80, 82, 117; Essex Deeds, I, 1; XI, 240; *Aspinwall Notarial Records* (vol. 32 of *Records Relating to the Early History of Boston*, Boston, 1903), 135–37.

which formed the major part of the land which was to become Hammersmith.[16]

When building operations began is not clear but, presumably, Leader's men went to work as soon as the five and a half acres of cornfield and the land needed for watercourse construction was in the Company's possession. There is little reason to doubt that Hammersmith dates from 1646, although, since it was conceived as a unit and developed as a complete integrated ironworks, as the archeological work of the Saugus Ironworks Restoration has made all but certain, it is quite likely that the work of building ran over into the following year. From a combination of archeological and documentary evidence we know that by 1650, when Leader quit his job with the Undertakers, Hammersmith consisted of a blast furnace, two fineries, a chafery, a big machine hammer, a slitting mill, and a smith's forge. Its water-power system was quite sophisticated. Water, stored in a large pond created by throwing a heavy earth and stone dam across Saugus River, ran through a sixteen-hundred-foot canal to a central reservoir, which in turn was tapped through wooden flumes that ran to furnace, forge, and slitting mill to provide the power for no less than seven water wheels. Although it was planned as a unit, Hammersmith, except for the basic water system, was in all probability built one piece at a time.

The stages can be dated only approximately. One reasonably good index to active ironworks operations in the early colonial period is to be found in grants of authority to dispense liquor to workmen. Ironmaking needed good lubrication. Here, however, no such grants are to be found. The Lynn local records are lost, and colony and county records are silent on the matter. Similarly, the appearance of ironworkers in the courts, and ironmakers tended to turn up there quite frequently, sometimes furnishes clues as to when work was going on and employees were drawing wages. The first mention of Hammersmith workers in the Essex Court records comes only in December, 1647.[17] Since something must clearly have been happening at the new site during 1646, we must suppose either that the workmen were still full of the fear of God and the Puritans who served Him, or enormously busy, unless the local authorities, anxious

16 *Essex R. & F.*, VII, 197–98.
17 *Ibid.*, I, 130.

to see the enterprise developed, were looking the other way from time to time. In neither the *Winthrop Papers,* nor in Winthrop's *History,* are any data on work in progress at Lynn ironworks available. Two measures passed by the General Court in 1646 carry no geographical references but may well suggest that work was well along. One was the award of a patent to Joseph Jenks, who, to judge from surviving documentary evidence, at least, had no connection with the Company's Braintree activities. Significant for its own sake as the first industrial patent granted in English America, it also suggests that, by May of 1646, there had been enough progress at the new ironworks to set one of its employees to dreaming of the prospects of setting up "engins of mils to go by water" for the making of scythes and other edged tools. The realization of Jenks' plans, as we shall shortly see, came more than a year and a half later. The other General Court action saw Leader getting "the countryes defective guns" at a price to be set by two Colony officials, presumably for resmelting in a blast furnace.[18] This, too, is a matter of some general significance as the first recorded use of "scrap" in metal processing in the English colonies. These may have been destined for smelting at Braintree Furnace but it is not unlikely that Leader was bent on using them, in conjunction with the bog ore, at his new furnace at Lynn.

That the new furnace was well along by 1647 is clearly suggested by Robert Child's March letter indicating plans to abandon Braintree Furnace "after another blowing."[19] It is all but inconceivable that such a step would have been taken had not alternative smelting facilities been presently or prospectively available. That both furnace and forge were built by at least January of 1648 is indicated in an agreement between Leader and Jenks which made it possible for the latter to set up his toolmaking plant. Here Jenks was authorized to set up a "mill or hamer for the making of sithes or any other Iron ware by water at the tail of the furnace & to have the full benefit of the furnace water when the furnace goes. . . ."[20] Leader was to supply him with "barr iron & cast iron for gudgins, shafts and hoopes." While in theory the bar iron could have come from Braintree Forge, it seems wholly unlikely that Jenks would have proposed to set up a

---

[18] *Mass. Records,* II, 129; III, 65.
[19] *Winthrop Papers,* V, 140.
[20] Essex Deeds, I, 79.

mill to make edged tools had he not been able to count on a ready supply of the wrought iron from which alone they could be made. Since he was allowed only six months to get his plant into operation, he could hardly have been starting work with an eye on remote future potentials.

Our first "production figures," however, date only from 1648. In August of that year, Governor Winthrop was able to report to his son, then at Pequot,

The Iron Worke goeth on with more hope. It yields now about 7 tunn per weeke, but it is most out of that brown earth which lyes under the bogge mine. they tried another mine, and after 24 hours they had a some (sowe?) of about 500: which when they brake they conceived to be a 5th par(t) silver; . . .

A month later the Governor had cause to pass along another and still more encouraging message, "the Furnace runnes 8 tun per weeke, and their barre Iron is as good as Spanish. . . ."[21]

For the following year, thanks to the survival of what appears to be either the first book of accounts kept by Leader's clerk, Thaddeus Riddan, or the "waste book" from which he copied his expenditures into the Company's permanent local records, we have something approximating real ironworks business data.[22] Lacking numerous pages, torn in many places and illegible in others, and all in all much mutilated, it is almost our sole guide to fiscal matters during Richard Leader's term of management. Unfortunately, also, its entries cover payments to workmen, almost without exception. Thus while we learn something of the number and status of the Company's employees, we are uninformed as to cost of purchase of raw materials, of building activities, and of the return on sales of the ironworks' products.

According to the Riddan account book, fourteen people, including a woman receiving sums due a deceased relative, were paid during 1649, either in cash or in bills drawn on local merchants, for work performed, for poll taxes and church assessments, and, more rarely, for equipment. Eight of these were being reimbursed for payments

[21] *Winthrop Papers*, V, 246, 262.

[22] The Riddan Account Book is now in the possession of Essex Institute. Interlineations, children's scrawlings, and the ravages of time make its effective use difficult.

to thirteen others, presumably their helpers, usually as wages but occasionally for clothing. Somewhat strangely, in both the "primary" and "secondary" accounts, the names of farmers and other local settlers outnumber those of people clearly identifiable as professional ironworkers. Of fourteen primary accounts, only three belong to regular iron craftsmen: Robert Ingalls, Joseph Jenks, and one whose name is missing but whose functions can be deduced from the kind of payment indicated for him. Seven were settled residents in the neighborhood—William Lakin, a Reading farmer; George Taylor, who was living at Lynn in 1635; Robert Parsons, a freeman of Lynn in 1635; Clement Coldam, miller and member of the Artillery Company in 1645; and Richard Mower, John Ramsdell, and Edward Baker, all of whom were established at Lynn long before ironworks had been thought of.[23] The four remaining are obscure. Matthew Boomer and Richard Barwicke are lost to history; one name is illegible; and the identity of the deceased relative of Ann Ingalls is not clear. Of the people receiving secondary payments, only five were known ironworkers—Henry Leonard, John Chackswell, Thomas Billington, Joseph Jenks, Jr., and a man whose family name is missing but whose service, "mending a pot," seems to put him in this category. The rest, unless farmers had sent sons of the same name to learn a trade at the ironworks, were all regular settlers.

While the entries spread over much of 1649, the total sums of the various accounts, usually indicated as "given in to Mr. Leader," are not large. With the exception of £24.4s. paid the man whose name is missing, and of which more than £23 covered the purchase of a house and lot, Joseph Jenks' £13.8s.3d. is the largest sum involved. Robert Ingalls got £5.16s.1d. and Richard Barwicke but 5s.6d. The rest average about £2, ranging from Lakin's £1.15s. to Ann Ingalls' £3.10s. In the secondary category, where we are dealing with individual payments and not with totals, entries of the following types appear: 14s. for a hat for Henry Leonard, and 10d. for gloves for Thomas Billington; £2.10s. for a suit of clothes for William Wilson; £1.3d. to John Chackswell, presumably as wages; and 4s. to Thomas Hartshorne, presumably for helping Lakin to carry wood. Unfortunately, the nature of the work performed is only very rarely specified. Making all allowance for the ravages of time, the conclu-

[23] The identifications are those indicated in Savage's *Genealogical Dictionary*.

sion must be that Thaddeus Riddan had kept less than the "perfect" accounts for which his master had established a reputation.

That there were more workers than these is clear from both the obvious lacunae in the Riddan account book as it has come to us, and the records of Essex County, where we find other names identifiable as ironworkers. There was old John Turner, who first appears as a man who had stabbed a daughter-in-law, threatened to kill a man, and been "overtaken in drink." Henry Stiche had broken the head of one Richard Bayly, as Richard Stiche and Nicholas Pinnion testified. All were ironworkers. Richard Pray was fined for swearing and cursing, for beating his wife, and for contempt of court, and the first three offenses had occurred at Hammersmith. Richard Greene, who offered testimony in a number of cases, was in all likelihood another of Leader's employees. John Vinton and John Gorum fall in the same category, as do John Dimond, Tobias Saunders, and John Smith.[24] Still others appear in contexts suggestive of ironworks associations, and, since there is reason to believe that not all of even the rough and rugged workers who staffed Hammersmith would have got before the courts, there must have been additional people toiling at the young ironworks. We have no basis, however, for even a crude estimate as to the total number of Leader's employees.

How these people lived, except for the negative that got accentuated in the court sessions, is also obscure. The workers and their families probably lived in simple, thatch-roofed cottages. Incidental testimony in some of the court cases carried reference to at least two houses in the occupancy of known ironworkers, and in one of them still another worker, a bachelor apparently, was getting board and lodging. Whether these were owned by the workmen or by the Company is not clear.[25] Certainly, these people could not have been accommodated in the one house mentioned in the inventory taken in 1650, when Leader handed over the Company's property to his successor. That contained only ten carbines, two tables, and two bedsteads!

---

[24] *Essex R. & F.*, I, 130, 136, 138, 156, 173, 174, 192, 198–99.

[25] *Ibid.*, I, 134, 136, 198; VIII, 200–203. An item in the Riddan accounts indicates a payment of £23.12s.6d. to one John Peperday for a house and lot, presumably for the worker under whose name the entry appeared. By 1653 several houses designated by the names of their occupants were carried as assets of the Company. There is only the Peperday item to suggest that "company housing" may have prevailed from the beginning.

Clearly, though, the workers had their own garden plots. So much we learn from a deposition that Nicholas Pinnion had come "out of his corn" and been heard to swear that "all his pumpions were turned to squashes, and by God's blood he had but one pompion of all."[26] Subsequently, as we shall see, the provision of Company-owned workers' housing became standard, and Hammersmith emerged as an "ironworks plantation" having much in common with what has later come to be known as the "company town."

If Leader lived in the house listed in the 1650 inventory, it could obviously have been only under rather less than comfortable circumstances. In all probability, however, both his personal inclination and the need to supervise work at both Braintree and Lynn led him to maintain his regular residence at Boston. Resident there at least as early as March, 1647, when Doctor Child was given the opportunity to be confined to Leader's house, on security of £800 under triple surety not to leave Boston town limits, he probably stayed in the capital for as long as he lived in Massachusetts. His seems emphatically *not* to have been company housing. While he came here assured that the Undertakers would give him living quarters, they doubtless had no intention in the world of setting up their agent in anything like the establishment he sold in 1655. Here was a "Mansion house . . . togither with the Orchard gardens Tymber yeards wharfes wayes water courses Grounds with all priviledges and Appurtenances to the same belongong . . ." which brought him £200.[27] The date of its purchase by Leader is not known. If this was the house in which Child was offered asylum, and in which Leader lived as agent of the Company of Undertakers, he had apparently been happily situated, both with respect to getting to his brace of ironworks, and to enjoying life in surroundings worthy of his station.

It seems clear that, despite Leader's efforts in negotiations with the magistrates, in complicated land purchasing, and in pushing construction and active operation at Lynn and Braintree, his ironworks got off to less than a flying start. In the early summer of 1647, Child had written Winthrop, "our Iron works as yet bring us in noe considerable profit. . . ."[28] Even as the Governor was reporting to his

26 *Ibid.*, I, 173, 295.
27 *Suffolk Deeds,* II, 210.
28 *Winthrop Papers,* V, 160.

son that things were developing more hopefully, Leader himself was indicating that he was plagued with troubles. On August 21st, 1648, he wrote the younger Winthrop:

The Company are very much discontented; and use me not as I have deserved. they have sent over one to take an account of things; and to give them sattisfaction how things stand with vs; But I am in some doubt they will be failed of their expectation in him. For my parte I am resolved they shall provide them another Agent, except a more cleere vnderstanding cann be mainetained betwext vs. . . .[29]

A month later, when Governor Winthrop described the "barre Iron" as "as good as Spanish," he also informed his son that, "the Adventurers in Engld. sent over one mr. Dawes, to ouersee mr. Leader for etc: they could not agree, so he is returned by Teneriffe."[30]

None of the correspondence between Leader and the Undertakers has survived, but it is not hard to tell what was basic to their dissatisfaction—delay in turning a profit on what had been a slow and painful start of their industrial enterprise. At one point, in their letter thanking Winthrop for his help in "the regulacion of some of our vnruly men," they seem to have been stressing technological difficulties. Here they wrote:

"Every new vndertaking hath its difficulty: Ours hath met with much: Casuall accidents have cost us very deare. And want of experience in the Mineralls in most of our workmen hath bin loss, and charge to us: . . . ."[31]

There is some reason to believe that making iron in the indirect process had been hard to achieve, although one must wonder what particular difficulties were present in handling bog ore, with which at least some of the workmen ought to have been familiar. William White, a man of some skill in science, wrote Governor Winthrop, as he sailed for Bermuda after arrangements to work for Child had fallen through, that had he come over with Winthrop the Younger, the iron ores of New England would have been tried with less cost by the "bloome harth" which he had used in Derbyshire.[32] Interestingly

[29] *Ibid.,* V, 248.
[30] *Ibid.,* V, 262.
[31] *Ibid.,* V, 209.
[32] *Ibid.,* V, 239.

enough, much of the later exploitation of Massachusetts iron was put through with the direct process, as White was apparently suggesting should have been the case from the start.

To judge from the tonnage figures cited by Governor Winthrop, however, the technological problems had ultimately been solved. This suggests that the Undertakers' grievances centered in the business side of the picture. It may be taken for granted that Leader had had to ask for more and more money, both fixed and operating capital. In theory, of course, the sale of ironworks products should have put an end to the call for additional funds to keep the plant going. Actually, however, the Undertakers seem to have insisted on being paid for their iron in money, and money was scarce in Massachusetts. The magistrates, at one point in the negotiations for ironworks concessions and privileges, spelled out the local situation:

. . . We acknouledge wth yow, that such a staple comodity as iron is agreat meanes to inrich the place where it is, both by furnishing this place with that comodity at reasonable rates, & by bringing in other necessary comodityes in exchange of iron exported; but as wee vse to say, if a man lives where an oxe is worth but 12d, yett it is nevr the cheaper to him who cannot gett 12d to buy one, so if your iron may not be had heere wthout ready mony, what advantage will that be to vs, if wee have no mony to purchase it? Itt is true, some men have here Spanish mony sometimes, but little comes to our smiths hands, especially those of inland tounes; . . . what monyes our smiths cann gett yow maybe sure to have it before any other; but if wee must want iron so often as our mony failes, yow may easily judge if it were not better for vs to procure it from other places (by our corne and pipestaves, &c) then to depend on the coming in of mony, wch is neur so plentiful as to supply for that occacon. . . .[33]

Unless this letter persuaded the English merchant investors that there was something quite special in the Massachusetts situation, Leader would have had trouble disposing of the ironworks' products, however strong a local market he had available. Since stocking and operating his plants took money, he must obviously have had to call on his principals for continuing advances of capital even after the works had been set up.

Locally, we assume, Leader followed the three standard procedures

[33] *Mass. Records,* III, 92.

that were open to him, paying for his supplies by merchants' bills, in what was essentially a barter transaction, meeting his obligations by bills of exchange drawn on the Company in London, and falling back on the expedient of going into debt by mortgage of ironworks property. We find evidence of the first of these in the Riddan accounts. There is no surviving record of the second, which was much used by Leader's successor, as we shall see, but it may be taken for granted that when the Company's agent could not pay for his supplies with cash or with iron, local merchants would have accepted bills drawn on London. Finally, there is evidence of mortgaging. In the fall of 1650, Leader "sold" the slitting mill at Hammersmith to Captain William Hathorne for £10.[34] The low figure suggests that the instrument was one of mortgage rather than of sale. It also suggests that Leader's position must have been close to desperate.

From what we know transpired in the case of the agent and factor who succeeded Leader, and from what we can infer from the general tenor of their letters to the magistrates, the Undertakers were probably given to constant interference in the management of their Massachusetts affairs. Their having done so is understandable. They had seen their money imperiled under Winthrop. Their costly plants were not in the clear a full six years later. It would have been strange had they not concluded that something was wrong with management. The Company's sending over of Mr. Dawes to "oversee" Leader doubtless followed a long exchange of detailed letters in which the Undertakers spelled out what steps ought to be taken, and in which their agent tried to make clear the obstacles confronting him. Communication was slow, and probably less than efficient, since the frames of reference of the parties on the two sides of the Atlantic were quite different. The investors must often have taken Leader to task. He must have been much troubled by the hand that reached out too often and from too far. The product can only have been a disgruntled agent eager for a change of occupation.

On August 28th, 1650, three of the Undertakers wrote to Webb and Tyng, two of the local shareholders, that Leader "hath dismist the works by the Consent of the Company and is mynded to follow his other occasions. . . ."[35] One would give much to know the details of the severance. There is little question that Leader had been follow-

[34] Essex Deeds, I, 18.

ing "other occasions" while still in the Company's employ. At least one surviving document suggests that he was carrying on mercantile transactions on his own as early as December of the previous year.[36] Now, however, he made a trip to England. During it, he was apparently so well able to render an account of his stewardship as to succeed in interesting John Becx, by then the chief figure in the Company, in some of the new activities which he proposed to carry on in America. Surely this is enough to suggest that relations between the Undertakers and their agent, once they had had a chance to talk things over, were wholly amicable. By the end of February, Leader was back in Boston getting ready to embark on the building of an ambitious sawmill in Maine.

In the interim, however, the ex-agent's relations with the Massachusetts authorities had turned out to be anything but amicable. Once free of his responsibility to protect the ironworks' interests, he had apparently given vent to feelings that had probably long been smoldering. Certainly, he had spoken his mind during his transatlantic voyage, as two fellow passengers attested. In May, 1651, the General Court found him guilty of formidable offenses, and sentenced him to the heavy fine of £200, plus £50, costs of court. According to the verdict, Leader had "threatnd, & in high degree reproached & slaundrd, the Courts, magistrates, & gouerment of this common weale, & defamed the towne & church of Lin, also affronted & reproached the constable in the execution of his office. . . ." Details of Leader's actions are unavailable. The reference to Lynn town and church may mean that the trouble went back to the controversy concerning the ironworks' exemption from local taxation. Certainly, however, Leader's lapses from accepted behavior must have been grievous. He eventually was persuaded to acknowledge that he had broken the rules of Christianity, morality, and civility, and to admit that the Court had cause to proceed against him. He presumably paid what was the second heaviest fine levied to that date in Massachusetts. In the matter of the remarks delivered while at sea, he was more fortunate, the Court concluding, seemingly with some reluctance, that its jurisdiction did not extend to words "spoken neere about the midway betweene this & England!"[37]

35 Ironworks Papers, 21.
36 Suffolk Deeds, I, 110.
37 Mass. Records, III, 227, 257.

Whatever the immediate cause of this contretemps, it is clear that Leader was no man to live happily in a community committed to the enforcement of a strict social, religious, and intellectual conformity. It has been suggested that he was probably on Doctor Child's side in the troubles that saw that distinguished figure put in his place in no uncertain terms. Certainly, he was in no sympathy with the established churches of Massachusetts, for the very good reason that he was a religious liberal. We learn at least some of the story from a very surprising source, the works of two cranky sectaries, John Reeve and Lodowick Muggleton, who styled themselves "the two last Witnesses and Prophets of the Most High God." They relate:

> In the year 1653 there came a certain Man, a Merchant and a great traveller into many ports of the world, and he was a religious man, but had somewhat declined the outward Forms of Worship because he could find no Rest there. . . . His name was Richard Leader. It came to pass when he came out of New England being persecuted there because he would not submit to their Forms of Worship, and when he came into Old England again he heard there were two Prophets risen up (Reeve and Muggleton). . . . He became a true believer . . . and shewed kindness unto John Reeve all the Days of his life, likewise his brother George became a true believer . . . .[38]

The sect's tenets were peculiar in the extreme. Its members did not believe in prayer; they were Unitarians; and their astronomy, based on a literal interpretation of the Bible, was pre-Copernican. This last area they debated with Leader, later publishing their version of the debate in a pamphlet called "The Discourse of John Reeve and Richard Leader Merchant."[39] They claimed to have converted Leader to their view that the sun was only a few miles up in the sky, and that thereafter he became "very mighty in Wisdom and Knowledge." For all we know to the contrary, he continued in the faith until the day of his death.

All this came after not only Leader's connection with the ironworks but his work in Maine as well, luckily for him, one is tempted to add, since the views in question would have been received with about as much sympathy as those of a Jehovah's Witness in modern Spain.

[38] J. and I. Frost, eds., *The Works of J. Reeve and L. Muggleton* . . . (3 vols., London, 1832), "The Acts of the Witnesses," 58–61.
[39] Copy in the British Museum.

Though the liberalism in the background of his conversion to the beliefs of Reeve and Muggleton might well have gotten him into trouble, there is no evidence that it did. In 1651, Leader was in Maine, building the sawmill in what is now North Berwick, which worked with "nere 20 sawes at once," and gave the Great Works River its name.[40] To that area he had been welcomed with open arms, the town of Kittery granting him not only land but sole use of the river and "all such tymber as is not yet Impropriated to any towne or persons."[41] The following year he was a magistrate, and also served as agent for Maine in England. While there he took a flyer in a quite different field, arranging with David Selleck of Boston to transport two hundred women and three hundred men, all rogues and vagabonds, from Ireland to New England and Virginia for sale as indentured laborers.[42] By 1656 he had sold out his interest in the sawmill, and was prepared, after another trip to England to raise capital, to try a new activity in a new place, saltmaking in Barbados.

Leader's Barbados career was less than successful. John Winthrop entrusted him with the secret of a new method for making salt in a series of ponds into which salt water was pumped to be evaporated. In 1659 he had the facilities all set up, but bad weather ruined the project just as it promised well. An attempt to refine sugar also went badly. As this counterpart of Winthrop put it, his condition was "subjected to change, Sometime to abound & sometime to want." Bad health got joined to economic adversity. In 1660 he was resolved to take his wife off to still another place, this time to Ireland, because of a "weaknes and feeblenes in all my lymbes, being the dreggs of a desperate disease which we call the Belly-Ake, which is only restored by remove into the colder clymes as experience teacheth." In the same letter, as he thanked Winthrop for an extensive account of mineral matters in New England, he expressed a judgment that is quite significant, given the man who wrote it and the man who received it, "Another age may bring something to perfection, when the country shall be furnished with men of partes to manage that busines." It was Maine, not Ireland, to which he turned for his colder

40 Winthrop Papers, 2.39.

41 *Province and Court Records of Maine* (Portland, 1928–), I, 152.

42 John W. Blake, "Transportation from Ireland to America, 1653–60," *Irish Historical Studies,* III, (1942–43), 271 *et seq.*

climate. There he went in the spring of 1661. There he died. A few lines from his last letter to Winthrop might well serve as an epitaph:

I question not but you are fully informed of the great change and sudden alteration in England. If the tender hearted ones are deprived the liberty of their consciences, to serve their God in truth of heart, but must all be forced to fall downe to what shall be Established by a State, my face shall be turned away from ever having thought to see my native country while I live.[43]

Such was the man, inadequately mirrored in scanty surviving data, who finished, if he did not start, Braintree Forge, who presumably planned and guided the setting up of all of Hammersmith, who played an important role in the early development of Maine. That he was on the whole successful in his management of ironworks affairs, our main concern here, is indicated by the comparative infrequency with which his name appears in Massachusetts Court records. If to Winthrop must go the credit for the initial impetus to New England ironworks, to Richard Leader, scientist, merchant, and engineering genius, is due the credit for their really effective start in the little industrial community of Hammersmith. If Winthrop may be called "America's First Scientist," Leader has, thanks to what he wrought on the banks of the Saugus and Great Works rivers, good claim to the designation, "America's First Engineer."

[43] 2 *M. H. S. Proc.,* III, 193–97.

8

Hammersmith—the Management

IT WAS UNDER the agency of Richard Leader's successor, John Gifford, that the ironworks at Hammersmith and the subsidiary forge at Braintree reached their high point of technological efficiency—and their eventual business failure. Ironically, the Undertakers' tragedy was great boon to us who would unravel the intricacies of a business carried on three centuries ago. Thanks to voluminous litigation to which the failure led, we have a better picture of what transpired at the ironworks, in good times and bad, than any now possible for other industrial enterprises in English America in the seventeenth century.

Because the materials surviving in rich profusion derive from controversy between capitalists and their agent and factor, between plant and business managers and "commissioners" and "attorneys" of the capitalists, and between any and all and the Company's creditors, evaluation is difficult. What was clearly beyond the capacities of a full generation of contemporary judges and juries is hardly to be solved from this distance. Nevertheless, by trying to stand aside from ever-recurring questions of individual culpability we can come close to a detailed and well-rounded view of management, technology, and workers during 1650–1653, the last phase of operations under the Company of Undertakers and, with little question, the most productive period of the ironworks.

John Gifford, the Company's third agent, stormy petrel of Essex

County and litigant *par excellence,* differed from his predecessors, first of all, in having a definite record of ironmaking experience and having been sought out by the Undertakers on account of it.[1] A resident and probably a native of Gloucestershire, he had worked for his father, John Gifford, senior, at what had once been the King's ironworks in the Forest of Dean. Here was a large plant, consisting of no less than three furnaces and three forges, that had been confiscated from a Royalist who held the ironworks on Crown lease and been awarded, by Parliamentary ordinance of 1645, to Colonel Edward Massey, the defender of the City of Gloucester. Gifford worked the plant, first in partnership with, and then by lease from, Massey,[2] until Parliament ordered the destruction of all the ironworks in the Forest in consequence of their heavy consumption of timber for charcoal. Although Gifford protested mightily, and despite the fact that he had been a loyal supporter of the Parliamentary cause and had served, along with his son, in a cavalry regiment of his own raising, his furnaces and forges were demolished in March, 1650. The younger Gifford had doubtless had good experience in the operation of a splendid plant. Now he had a compelling reason to seek another connection —unemployment.

Unlike Leader, Gifford also had had at least some familiarity with life in the Bay Colony. Little is known of his first stay in Massachusetts. From that faithful reporter, Emmanuel Downing, writing to the younger Winthrop about still another change in the ironworks management, however, we learn: "Here is one Jeffries (Gifford) come in Mr. Ledders place. he was heretofore maior Gibbons man. he hath ben these 4 or 5 yeares past imployed in England as Clarke to an Iron worke."[3] Major Gibbons was Edward Gibbons, in 1649 major general of the Massachusetts militia group.[4] Gifford could hardly have been his "man" in the old country, since the Major had been at Mount Wollaston some years before Boston was settled, at Charlestown in 1630, and in Boston from 1631 on. A man of substance, with water mills at Boston and Braintree, he had probably employed Gifford as a young apprentice clerk.

[1] C9/21/5.

[2] Historical Manuscripts Commission, *Seventh Report* (London, 1879), 14; S. P. 23/136/211.

[3] Winthrop Papers, 2.39.

[4] Savage, *op. cit.,* II, 245.

Despite these qualifications, Gifford was hired by the Undertakers on a more restricted basis than had presumably prevailed for Leader, and at lower salary. His contract has survived. By its terms, nine members of the Company, Frost, Bond, Copley, Foley, Child, Becx, Pocock, Beeke, and Foote, authorized Gifford to proceed to New England, receive the Company's assets from Leader, and take over the management of the ironworks for seven years. The Undertakers were to pay his passage, provide him with suitable housing, and give him a salary of £80 per year. In return for this, plus assurance of a raise when he had demonstrated his "industry, care & honestey," Gifford was so to carry on the works as to bring the greatest possible benefit and profit to his employers. Besides the faithful care of the ironworks and their land and timber holdings, the new agent took responsibility for seeking out new mines of iron, lead, tin, silver, and other minerals.

The greater part of the document consists, however, of restrictions on Gifford's activities, and other devices aimed at protecting the Company's interests. The new agent was to be bound by his contract and such advice and orders as he would from time to time receive from the Undertakers, or any four of them. When he took over from Leader, he was to obtain a strict accounting of the Company's dealings and an inventory of its property, and send them to England. Further, inventories and accounts were to be sent to his principals annually or as often as the Undertakers should require. Gifford agreed not to erect ironworks of his own, and not to engage in the iron trade for others without the Company's consent, during *or after* the seven years of his contract, under penalty of having to pay £10,000 for every unexpired year of contract and for every year in which he might be so employed thereafter!

Even the guarantee of his housing was coupled with a proviso. Gifford was to get a "convenient house for his habitation" if his "perticular paynes and Extraordinary Expence of time in & about the said house" did not deprive the works of proper supervision. If Leader's effectiveness had suffered from his "other occasions," the English gentlemen were clearly seeing to it that their new man not be similarly tempted. Finally, as though all this were not more than sufficient, the Undertakers required that in "all Matters and businesses of weight and Speciall importance" Gifford should solicit the advice and help of such of their number as were or might become residents of Massachusetts, and also of John Endicott the Elder and William Hawthorne. He was to follow the advice of the "major part" of these local people, except where it ran counter to express instructions from the Company. In that event the problem was to be referred to London for settlement by the English shareholders or by a majority of them joined with a majority of their Massachusetts fellow investors.[5]

These control devices are the more impressive because, strict and cumbersome as they were, they applied to only one major phase of management of the Company's affairs. Another was carved off and placed in the hands of a "factor," William Awbrey, engaged under formal contract by the nine shareholders who had hired Gifford. While the Undertakers insisted that they had "multiplied their charges" in sending this London merchant to handle the business end of ironworks affairs, they claimed to have found it advisable for "2 to be ther in equall trust in case of mortallity as well as otherwise And that the Worke might haue beine so much the easier to be Well performed."[6] They might also have said, and with more candor, that they were now bent on following the old system of "Divide and rule."

By the terms of his seven-year contract, Awbrey was to go to Boston as soon as possible, hire a house and a convenient warehouse, and there receive and care for shipments of materials from the Undertakers and of finished products from the ironworks. These he was to sell on the best possible terms and keep the ironworks supplied with money and goods out of this stock and the profits on its sale. In his account books, which he was faithfully to keep "after the best way of keeping Merchants Accounts that he can," the advances of operating capital were to be handled as debits, and the iron goods he received, charged by quantity and quality at current prices, as credits. Reports, inventories, and accountings were to be sent to England as often as required. Good clerks and factors were apparently less scarce than good ironmasters. Awbrey, at any rate, was bound to serve for but one year, although at a salary only £10 lower than Gifford's.[7]

Little is known of Awbrey's background. The name is a fairly common one, and identification among several William Awbreys liv-

[5] Ironworks Papers, 13–20.
[6] Ibid., 31.
[7] Suffolk Deeds, I, 216–18.

ing in London in the 1630's and 1640's is all but impossible. The new factor must have been either a quite small merchant or a merchant's clerk. The complete absence of references to him in the English courts suggests that his operations had been unusually modest or unusually free from controversies. That in Massachusetts he became a man of some substance is suggested by the size of his salary, his eventual admission to the Artillery Company, and his marriage to a daughter of Edward Rawson, the Colony's secretary.[8] In at least one deed he is referred to as "Gentleman."[9]

At first glance the division of functions between Gifford and Awbrey seems to make sense. Gifford was to make the iron, to serve as what we today call a "production man." Awbrey was to be the Company's fiscal and sales agent. While the two men were to be in "equall trust," and while their respective spheres of operations were quite well delineated, their mutual working relationship was left unspecified, apparently, except for the requirement of an exchange of accounts in which Gifford would indicate what he had received by way of supplies, and Awbrey, what he had gotten in ironworks products. The looseness of arrangements may have been fine for a plan of "Divide and rule." It was less than good, given the fact that in certain areas, as in direct plant sales of iron, there was bound to be overlapping. It was even worse when one of the parties was that arch-independent, John Gifford.

Awbrey, though hired later, seems to have reached New England earlier than did Gifford, probably in the fall of 1650. When Gifford arrived, he received an inventory of the stock and tools at Hammersmith and Braintree taken in late December of that year and indicated as having been given into his hands by Awbrey and William Osborne.[10] The latter had been serving as agent for Richard Leader, who had long since gone about his own affairs. From as early as the previous May, perhaps, he had been acting as "agent of the ironworks for the time being."[11] Gifford thus took over, not from Leader but from the new factor and an agent pro tem. It is quite probable that more

[8] Savage, *op. cit.*, I, 78.

[9] *Suffolk Deeds*, I, 178.

[10] *Essex R. & F.*, I, 294–95.

[11] *Records of the Court of Assistants of the Colony of the Massachusetts Bay, 1630–1692* (3 vols., Boston, 1901–28), III, 23. The interim "Agent" was subsequently dismissed by Gifford. (Ironworks Papers, 34).

was involved here than Leader's absence. If we may believe some later testimony of Gifford's, at any rate, he had pressed Leader for the detailed accounting for which his own contract called, and his predecessor had not only flatly refused to deliver it, but would not let anyone else make it available.[12] The reasons for this state of affairs are obscure. So far as we know, there was no resentment between Leader and the Company. Presumably, at least, it would have been too early for Leader to have developed a personal animosity toward his successor. Gifford, however, had an eloquent tongue and a rare talent for sounding off. In his new situation, again to accept his word for it, he had found plenty to talk about—and all of it bad for Leader.

According to a later statement, Gifford, on his arrival, had found the ironworks in sorry shape indeed. He asserted that,

He did with all dilligence . . . apply himselfe to the management and this first by takeing all the care he could in the repayring of the said Workes which he found in a ruynated conditon, as the forge and furnace ready to sincke, the wheels and trowes broken, the Waste gates ready to drop downe and to be carryed away by the waters by which meanes the buildings were all in danger of being washed away, the which this defendant was forced to the setting downe and makeing most of them all newe, the works at that time not going, and the said works were without stocke either of Myne or Coales to sett them in worke and scarcely a man there to any Work but whose tymes were served and become free.[13]

Here we have evidence on the other side. The Undertakers themselves indicated that by 1650 "the worke of findinge and makeing iron came to very good perfeccon" and expressed hope that "in a few years" they would not only recover their original investment but turn a profit as well. Leader was reported to have said that he would give £1400 a year for the ironworks. A collier, Richard Smith, insisted that he knew the works had been fully developed before John Gifford arrived and could make iron as well before as after his taking up the management. Four men with direct contact with the ironworks bore him out, and another employe, John Vinton, announced that he considered the works worth £700 a year. And John Becx, writing in 1652, claimed that "in osbornes time theare was good store of barr Iron,

12 C9/21/5.
13 C9/22/24.

sow Iron, Potts, Coale, wood and myne leaft and the works in good repayre. . . ."[14] Where the truth lay we know not. Certainly, however, if Gifford went around Boston expressing opinicns like those he subsequently put on paper, Leader's refusal to have anything to do with him, even in a routine transfer of accounts and inventories, would have been understandable!

There is no question that the new agent, never a man to hide his light under a bushel, started work with great expectations. In 1651 he wrote the Undertakers that he hoped to make four hundred tons of bar iron a year. Somewhat later, perhaps early in the following year, he was a little less optimistic; still he was expressing the hope that he could make three hundred, "unlesse the workemen prove lazy." Since he was eventually to claim that for want of stock he had spent a full year before the furnace went into blast, it looks as though he had been mightily optimistic, or Leader had left things in better shape than he saw fit to admit, or both.[15] There is some evidence to support the notion that Leader's stewardship had not been poor. He had indicated that when he left the works, there "was Remaineing in stock and good debts aboue 200 (pounds)."[16] This, to judge from the 1650 inventory, was most conservative. Working tools were rather less numerous than those on hand in contemporary English iron-works, but there was more than £200 worth of ore alone, and the value of pig iron, cast products, bar iron, and charcoal at Hammersmith and Braintree came to well over £1900.[17] Trying to strike a balance among all these conflicting claims, one must at least wonder if Gifford had had much more to do, at the beginning, than attend to the acti-vation of plants which had ground to a slow down or even a complete halt in the interim between Leader's going and his own arrival.

Awbrey, we know, got off to a poor start. Doubtless set up for business in Boston during 1651, it was only in February of the follow-ing year that he there bought a house, wharf, and land for £145.[18] From a scorching letter addressed to Gifford by the Undertakers, we learn that Awbrey had been most remiss in explaining bills of ex-change which he had drawn on the Company. In one letter he had

[14] C9/21/5; C22/956/7; *Essex R. & F.,* II, 90.
[15] Ironworks Papers, 32, 37, 253.
[16] *Ibid.,* 33.
[17] *Essex R. & F.,* I, 294–95.
[18] *Suffolk Deeds,* I, 178–81.

pleaded that he was "troubled with melancholy." The English investors, claiming that his letter made enough sense to eliminate this possibility, announced that they feared "more knavery than Mellancolly," and indicated that once his accounts had been approved by certain "commissioners" being set up to oversee the Company's activities, he would be discharged.[19] If we can trust the data contained in a letter which is full of cold fury, discordances between Gifford's and Awbrey's accounts, and clashes between them, had already become apparent.

In whatever shape Gifford had found the works, he must have been a very busy man. In residence at Hammersmith, in the "Ironmaster's House," on which he had lavished costly repairs,[20] he had responsibility for the whole range of production activities, for the supervision of the workmen, and for the sale of part of the ironworks' products right on the spot. To him came letters from the Undertakers, now advising and instructing, now criticizing. Around him lived the members of the established Puritan community, with whom he had to keep on good terms and to whose standards he had somehow to make his workers conform. In his care, in other words, were planning and production, personnel and community relations, and plant accounting responsibilities. All involved rather more than is handled by present-day counterparts of this seventeenth-century ironworks agent, as will be clear in the following functional analysis of the operations over which he presided.

First came the wood cutting. This work was carried, during Gifford's tenure, by neighboring farmers and assorted craftsmen employed on a part-time basis, by skilled ironworkers who went into the woods when their own work was slack, and by Scots indentured

[19] Ironworks Papers, 32, 35.

[20] *Essex R. & F.*, II, 89; VIII, 201. The structure in question is presumably the present Ironmaster's House at Saugus. Its dating is difficult. The claims of local historians that it was built by Thomas Dexter in the 1630's are rendered more than a little suspect by the presence in its foundations of burned out sandstone, apparently discarded blast-furnace hearthstones which could only have become available after the furnace had been in blast. From Becx's comment that Gifford ought to have been content to live in the house as Leader left it, it is clear that the house was standing in Leader's time. It is equally clear that a house was included in the Dexter property bought by Leader. Since the 1653 inventory of the Company's estate mentioned the "dwelling house that Mr. Gifford lives in" *in addition to* the Dexter farmhouse, it would appear that the Ironmaster's House had been built during Leader's tenure as agent.

servants. The ironworks neighbors working in off season, and the ironworkers who from time to time took up this unskilled work, were paid two shillings a cord for cutting and six shillings a score for cording.[21] Since the outlay of cash was as unwelcome to the Company as it was highly convenient for farmers who saw little hard money, the Undertakers arranged in the latter part of 1650 to send to Massachusetts numbers of Scots taken prisoner in Cromwell's battles and sold to John Becx and Company for use as indentured servants. They had to be trained for the forest work, which was but second nature to the farmers of the towns surrounding the ironworks. This task fell to ironworks employees rather than to ironworks neighbors hired part-time.[22] It must have been well handled. During 1651 some forty-five men, exclusive of the Scots, drew 255 pounds and 18 shillings for work on cord wood. By the following year the Scots had so well taken hold that they carried far and away the greater part of the timber operation, and were credited in the accounts with having cut and corded more than four thousand cords of wood, at a gross saving to the Company of about £500.

Next came the conversion of vast quantities of wood into charcoal. The process will be described in detail below. Here we need only point out that the colliers and their helpers, living in huts in the woods, piled the wood in mounds, covered these with turf or leaves and earth, set them afire in a slow charring flame, and then stood guard endlessly to see to it that the operation was carried through safely and successfully. The work was highly skilled and so well rewarded that it was almost certainly the best paying job at the ironworks. The Undertakers expressed amazement at a going rate of £27 for a hundred loads of charcoal. A skilled finer announced that he could earn more and with less output of energy as a collier than he could at his own highly specialized job. During 1653 the colliers were paid five shillings and five shillings, six pence a load, with one man earning more for coaling than Gifford did as manager, and two others doing almost as well. Surviving charcoal accounts are fragmentary. In them, we find that the making and hauling of charcoal cost the Company more than £1100 during 1651 and 1652, although total

---

[21] Ironworks Papers, 77–84. Cording was the piling up of wood in cords. The score was twenty cords.

[22] *Ibid.,* 77–80.

production is not indicated. In 1653, however, seven colliers, apparently working at separate locations, turned out more than fifteen hundred loads of coal.

John Gifford would have had little, if any, occasion to supervise either of these forest operations directly. In the cutting and cording, certain men received payments for "setting on" wood cutters, presumably as foremen of some sort. Piece-rate wages guaranteed good performance by the workers. There was some danger of fire, and on one occasion a considerable stock of cord wood was destroyed.[23] Given the huge store of timber available, there was doubtless little or no need for Gifford to worry about the preservation of resources and the reforestation which figured largely in English operations of the period. The colliers ran their own show. The agent, however, had responsibility for seeing to it that the loads of charcoal were full measure as delivered at the ironworks for storage at the "coal house." This was not always easy to gauge, but it was subsequently alleged that Gifford had been remiss in the matter.[24] In precautions against fire, another area of managerial responsibility, Gifford seems also to have fallen down. In the eyes of the Company's local representatives, at least, it was his dereliction that permitted the breaking out of a fire which at one point destroyed a large quantity of the precious ironworks fuel.

As men worked to make available the charcoal which went into the maws of the blast furnace and the forge hearths in tremendous quantities, others were busy in finding, digging, and hauling the bog ore. These operations were carried on over a wide area in Gifford's time, ranging from as close as the immediately adjacent farm of Adam Hawkes, and other Lynn points, to Reading and even as far as Hingham.[25] It is doubtful if ore-prospecting got carried on systematically, although it is impossible to tell whether one man who got paid £1.10s. for "findeing a parcell of Bogg myne in the woodes" had come upon the deposit by accident or by design.[26] The actual mining

[23] *Ibid.*, 40, 43, 46, 77, 78, 80, 83, 85, 124.

[24] *Ibid.*, 190.

[25] According to the Lynn local historians, the bog ore on the Hawkes land had been the prime stimulus to that town's "first" ironworks. While the taking of ore from the the Hawkes land is attested in surviving portions of the Company's accounts, it is also indicated that ore came from more distant points as well. Topographical change and vagueness of citation in the documents make precise identification nearly impossible.

[26] Ironworks Papers, 122.

was a matter of taking out, with pick and shovel, masses of limonite, some of it of earthy consistency, some of quite solid chunks, in swamps or low-lying areas which had once been the beds of ponds or stagnant streams. Part of the ore was obtained from lake and pond bottoms by a backbreaking manipulation of long-handled scoops or "floating shovels." Mining was carried on by both part-time and regular ironworks employees at rates ranging from a shilling to three shillings a ton. Once mined, the ore had to be hauled to the plant. This was done by local residents, who were well remunerated. In 1651 the standard rate seems to have been six shillings the load. In 1653 hauling charges ran between six and seven shillings, the higher rates apparently reflecting carriage of ore over greater distances. Neither mining nor carting took many hands. Payments were made to no more than fifteen men in the full span covered by surviving accounts. Total ore costs ran high. The mining and carriage of one thousand tons of bog ore cost £305 during 1651. Two years later only half as much was handled, and the bill came to more than £200.[27] Managerial responsibilities probably involved no more than checking on weight and quality. In all likelihood, however, the agent had also to work out the terms on which the owners of the land from which the ore came were paid.

Besides fuel and ore, the furnace operation took some form of fluxing agent. At Hammersmith, a gabbro mined at Nahant, and carried in the accounts in a category labelled "rock myne," filled this bill. Stray documentary evidence and the findings from archeological work carried on at the plant site suggest that minor quantities of imported limestone and coral had been used as flux.[28] Both the form of the raw materials accounts and recent laboratory experiments make it all but certain, however, that what men had probably first turned to as a source of iron was normally used to rid the bog ore of its impurities. The mining of this dense igneous rock right at the ocean side was far harder than that of bog ore. It took sweat to pry the rock loose by drilling and heating, to burn or roast it to get it into workable size and eliminate some of its impurities. The work, handled by regular ironworks employees, nevertheless brought the same pay as did that on bog ore, a shilling a ton in 1651, a shilling

[27] *Ibid.*, 87–88, 122–23; *Essex R. & F.*, II, 94.
[28] Ironworks Papers, 142.

and a few pence in 1653. Hauling, by cart rather than by boat, cost 1s.4d. Quantities handled were large—five hundred tons in 1651, nearly two hundred fifty in 1653. Total costs, covering both mining and carriage, ran to about £66 and £29 in those years.[29]

The various materials, received at the ironworks under what one hopes was the watchful eye of John Gifford, had to be stocked against the time when all would be ready for a blast. The bog ore was probably given some form of crude shelter. The gabbro could hold its own against the elements. The charcoal went into a stone-walled structure of crude but solid rubble masonry probably roofed with planks. Located far enough from the furnace to be beyond reach of its sparks but convenient to it and the forges, it was valued in a 1653 inventory at only £15.[30] There were also ore-roasting facilities close to the furnace. When all was set, the ironworkers who had joined the part-time employees in the gathering in of the raw materials returned from swamp and forest, from the colliers' pits and the quarry at Nahant, ready to resume their proper professional skills. Now, the actual work of making iron could begin.

Over a wooden bridge running from a natural elevation near the agent's house to the furnace top passed a steady procession of furnace loaders. Into the charge hole went basketful upon basketful of charcoal, flux, and ore, layer by layer. Water-wheel-driven bellows were set in motion. The furnace was lit and, as the fires became stronger and stronger, the smelting process began. At the furnace base the founder and his helpers carefully checked its progress, now adjusting the blast, now preparing the V-shaped trenches into which the liquid iron would be tapped and allowed to cool as sows, now burying and getting ready the clay molds used in the pouring of pots and other hollow ware. At appropriate intervals they tapped the furnace. The over-all job, however, continued night and day for as long as the stock lasted, or until cold weather froze the wheel, or the crucible lining burned out.

Much of this was obviously highly skilled work. It is not at all clear whether Gifford had to give it close supervision. The founder was in direct charge, and it may be wondered if the agent did more than

---

[29] *Ibid.,* 89, 122, 124. The sums of the various payments listed work out to the totals indicated in the text.

[30] *Essex R. & F.,* VII, 201.

check on the cost of operations and the amount that they produced. He or his clerk, or both, doubtless kept records of fuel and ore consumption. They were responsible for making and recording payments of wages in what appears to have been a somewhat complicated system. The founder's wages were apparently figured by the week. So were those of the fillers, the men who kept the furnace charged. General helpers were paid by the day and other workers who had put through designated jobs, chiefly matters of maintenance, by the piece. Repair and upkeep were extensive and costly. Gifford's accounts for 1653, for example, show payments for such jobs as replacing the wheel shaft, putting in a new hearth, and fixing up the furnace wheel and bellows.[31] Counting both maintenance and production chores, we find no less than nine workers in the furnace area besides the founder and the potter, a man on whose activities the accounts are wholly silent. Most of these, however, were carrying on odd jobs and ought not to be classified as true furnace workers. The total volume of furnace activity, even as mirrored in less than good data, is, however, quite impressive.

Most of the furnace output came in the form of cast-iron sows, which were hauled to the nearby forge or shipped to that at Braintree to be oxidized into wrought iron, the works' chief article of sale. Hammersmith Forge consisted of two fineries and a chafery set up around a single power hammer. Here strong arms brought up a sow and stuck it through a hole in the wall of a finery, a masonry structure capped with a "larthed and daubed" chimney, and rather like an overgrown blacksmith's forge. In this position the sow slowly melted down, drop by drop, in a charcoal fire burning under blast in the finery's iron-lined crucible. The newly molten iron which gradually accumulated in the crucible was worked up into the shape of a pasty mass, the "loop," taken out, and, first under hand hammers and sledges, then under the blows of the big power hammer, "shingled," or given the shape and consistency of a "bloom." This, in turn, got split and each half carried, by repeated heating and hammering, through the "anchony," a bar with rough ends vaguely like a dumbbell, and thence to the finished "bar." The heat for drawing out the larger nob on the anchony was provided by the finery, that for the smaller, by the chafery, a somewhat larger but quite similar unit.

[31] Ironworks Papers, 62–64, 114–21.

Since the finery carried all of the roughing-out work and some of the finishing, the chafery, only the remainder of the finishing, one of the latter could easily keep up with two of the former.

The outline of the process hardly does justice to the drama and color, the skill and timing, of the work involved. The material was hot iron, dragged back and forth between hearths and power hammer over iron plates on the forge floor. The impurities that were being squeezed out as the big hammer banged away flew through the air as sparks or trickled down to the anvil base to cool as scale. The clatter of tongs and other hand tools, the whish of the bellows, the rhythmical thud of the power hammer raised by cams on a wheel shaft and lowered under impact from a wooden spring as well as by its own great weight, the voices calling, now for more blast, now for faster hammer action —all this made up a cacophony as rich and varied as the quick changes of light and color in the rather dark forge building. Here, truly, was the working area of specialists.

The working complement of Hammersmith Forge in Gifford's time seems to have been on the order of ten or a dozen men, most of them skilled in the extreme. As with the furnace specialists, it is difficult to tell to what extent the agent left them on their own. Certainly, he kept watch on forge-operating costs, the efficiency of the process as measured by the loss of weight in the conversion from cast to wrought iron, the quality of the finished product, and production tonnages. Given the nature of the actual work, however, it is more than doubtful if Gifford offered more than infrequent and casual overseeing of tasks about which the workmen were doubtless far better informed than he, however much experience he had had with ironworks.

Two members of the forge crew ran the fineries, one the chafery and hammer. The jobs of the others are not wholly clear. Several must have been helpers to these master workmen. The making and drawing out of iron was reimbursed by the piece, the helpers' general work by the day, and that involved in upkeep and repair at so much per given job, although the wage system seems to have been even more involved than that which prevailed at the furnace. As in that branch of the Company's activities, maintenance took a lot of attention. Here expenditures were regularly made for repairing wheels, daubing chimneys, dressing bellows, erecting new flumes, and so on.[32] Given the

[32] *Ibid.,* 114–15, 119, 120.

heavy fixed capital investment concentrated in the forge, and the centrality of its product in the whole operation, its care must have loomed large in the agent's mind, even if its real work were left pretty much in the hands of the men who labored there.

Some of the bar iron was carried through still another process, that of the rolling and slitting mill. Here was a rather complicated machine operation which saw the bar iron heated in a reverberatory furnace and passed through a pair of power-driven rollers for flattening, and a set of opposing disk cutters for slitting into rods. The rod iron, stacked in bundles, was sold mainly for use in the making of nails. This mill had no special crew of its own. When rod iron was to be made, workers who carried on the work of the forge put aside their ironmaking tasks and took up what was really a fabricating operation. During 1651 seven men were employed at the slitting mill and an eighth in hauling off some of the finished products. Wage data for 1652, as for the other plant units, are not clear. In 1653 only three men seem to have been paid for work in the slitting mill.[33] Payment was both by the day and by the piece, with items of the latter type covering maintenance and repair work. Whether because day rates on the routine work failed to bring forth maximum efforts or because the work was unusually strenuous, the slitting-mill hands seem to have received a somewhat conspicuous lubrication. We have record, at any rate, of two allowances of cash for workmen's drinks in a fairly brief interval.[34] How much supervision the slitting-mill operation demanded is nowhere indicated.

In all this, Hammersmith has been our focus. Gifford, of course, also had the ironworks operations at Braintree to worry about. There woodcutting, coaling, the mining of bog ore, and finery and chafery activities were being carried on. Between the main plant and its subsidiary traveled small craft laden with sows from Hammersmith Furnace to be reduced to wrought iron at Braintree Forge, and, on the return voyage, with ore from Hingham and Weymouth which would help to feed the furnace at Lynn. There was also quite a lot of exchange of workmen between the two places. The techniques in use were the same at both plants; so were the wages paid. The work crew at Braintree Forge was smaller than that at Hammersmith.

[33] *Ibid.*, 75–76, 114–15.
[34] *Ibid.*, 76.

In both places, however, there were both part-time, farmer-neighbor workers and full-time skilled specialists. There is no need here to analyze the Braintree operation in detail. The branch plant was essentially a Hammersmith in miniature, minus furnace and slitting mill. Its management posed some problems but these went back mainly to distance. Setting up a branch manager would have made sense, and the possibility seems to have occurred to the Undertakers.[35] Apparently, however, it was never put into effect, and through his whole term, Gifford ran the whole ironworks show, going to Braintree with his clerk from time to time to handle the payment of workers and other concerns of the Monatiquot River plant.

Besides managing all these production activities, Gifford had the care of a six-hundred-acre farm at Hammersmith. Work in this case involved the cutting of firewood for the manager's and workers' houses, the care of cattle, the growing of hay, and, in all likelihood, the raising of vegetables. While a number of regular and part-time ironworks employees were paid for specific jobs like woodcutting, grass mowing, and the making of household furniture,[36] it looks as though the bulk of the farm labor was carried by the indentured Scots. It is doubtful if the farm ever filled its role in building up the self-sufficient industrial community which the Undertakers seem to have envisaged. It did give rise to controversy. Gifford was eventually accused of having put some of the farm and some of the Scots labor to his own use. He had a team of his own which was used in hauling ore and other ironworks supplies and paid for at the same high rates that prevailed for neighboring farmers, who fared well indeed in carting for the Company. Since his oxen were grazed on Company land, it is possible that the ironworks farm was a better thing for the agent than for his employers.[37]

John Gifford was also responsible for the Company's indentured servants, both English and Scots. They had to be fed, clothed, and provided with medical care, tobacco, and other incidentals. The bills for all this ran high. During 1651 twelve indentured servants, some of them, at least, recruited in England, worked at Hammersmith or Braintree, or both. Their board, taken in the homes of ironworkers,

35 *Ibid.*, 34; *Essex R. & F.*, II, 88.
36 Ironworks Papers, 108–13; *Essex R. & F.*, II, 91.
37 Ironworks Papers, 120, 125; *Essex R. & F.*, II, 91.

cost five shillings a week. Their clothing was made, now by local tailors, now by the wives of ironworkers, of materials bought from Boston and Salem merchants. Some were farmed out to skilled workers, but Gifford was indirectly as responsible for them as he was directly for those who stayed on as servants of the Company proper.

The Scots posed special problems. Here were not only ironworks laborers, but workers brought over as a business investment, their services destined to be sold to whoever would buy them. Only thirty-five of a total of sixty-one carried on the books in 1651 stayed on as an ironworks unskilled-labor reservoir. Of these, seven had been farmed out to ironworkers; the rest belonged to the Company.[38] Though all the others had been sold off, they had been, for longer or shorter intervals, Gifford's charges. The Scotsmen's diet seems to have amounted to slightly more than five shillings a week. Other costs ran higher than in the case of the English servants. Housing had also to be provided. Some were quartered in workers' cottages. Most were sheltered in a house newly built for them.[39] In wretched shape when they came off the boat, it took time and money to bring them to health. Few spoke English, and their adjustment must have been as painful to the agent who had them in his care as it was to themselves. One index of the success of that adjustment we have already encountered in their timber-cutting job during 1652. Others aplenty will appear below.

The Undertakers' agent also bore the responsibility of a small fleet of vessels. While land carriage was handled by rented teams, Gifford's included, the Company owned, by 1653, two boats of moderate size, a small boat and a skiff. One of the former was referred to as "the Company's boat" and was apparently the prinicipal means of coastwise and river shipping, not only between Hammersmith and Braintree, but between both plants and Awbrey's Boston warehouse as well. In the period 1651–1653 six men were paid wages for "several voyages"—to Boston, Braintree, Weymouth, and Hingham—and for various ship repairs. The sums involved suggest that the voyages were

[38] Ironworks Papers, 97–107.

[39] This was presumably the present Bennett-Boardman House in Saugus or, according to claims advanced by Abbott Lowell Cummings in "The 'Scotch'-Boardman House, A Fresh Appraisal," *Old-Time New England,* XLIII (1953), 57–73, an earlier structure replaced by it, almost on the same site and prior to 1700.

by no means infrequent. In 1651, for example, Theophilus Bayly and Ephraim Howe got £25.5s. apiece for "going with the boat." In 1653 the Graves brothers jointly received £26 and Bayly and John Lambert £28 for their "several voyages."[40] Data on the boats themselves are not available.

While the sale of iron was presumably a responsibility of Awbrey's, it is clear that Gifford sold some iron at the plant. In his accounts covering 1652 and, apparently, part of 1653, sales of bar iron totaling 12 T. 13 cw. 3 qr. 23 lbs. and worth £253.13s.10d. are itemized.[41] This by no means large quantity was divided among forty-five purchasers ranging from Governor Endicott to Boston and Salem merchants to various full-and-part-time workers at Hammersmith. What went to the workers was slight indeed, although the combined purchases of Joseph Jenks, now presumably operating his own toolmaking plant, made him the ironworks' second-best, plant-sales "customer." Scattered in various court depositions, too, are references to purchases "at the works." By inference from the place of residence of several of the buyers mentioned, it seems clear that this meant both Hammersmith and Braintree.[42] Obviously, furthermore, Awbrey was selling iron from the warehouse. None of his business records survives. We know, however, to cite but a single piece of circumstantial evidence, that Gifford sent him between December, 1650, and June, 1652, 128 T. 3 cw. 2 qr. 27 lbs. of bar iron.[43] Letters from the Company make it plain that he had not sent anything like this stock of iron to England! In no case, unfortunately, is a clear index of total sales or of ratio of plant to warehouse sales available.[44]

It was clear in the terms of Gifford's and Awbrey's hiring that the setting up of sound accounting procedures had been much in the Undertakers' minds. The general accounting system is reasonably plain in surviving ironworks documents. It has been apparent in our analysis of plant operations that Gifford kept a record of expenditures set up in categories—general outlays, furnace, forge, slitting mill,

[40] Ironworks Papers, 56–57, 119.

[41] Ibid., 48–51.

[42] E.g., Essex R. & F., II, 30, 70–72, 91–93, 98; IV, 391, 393, 421.

[43] Ironworks Papers, 249, gives 128 T. 10 cw. 2 qr. 27 lbs. The figures cited in the text are those of a stray fragment in "Miscellaneous Manuscripts" (M. H. S.), and check with Gifford's proper accounts.

[44] Ironworks Papers, 22–24, 31, 33, 142, 145, 287; Essex R. & F., IV, 393.

flood gates, cordwood, charcoal, bog ore, rock ore, English indentured servants, Scots, and the farm.[45] Expenditures at Braintree were carried, under appropriate heading but with rather loose designation, in these general cost accounts. Each payment to a particular individual was keyed, when the accounts were well kept, to the page in the Company's account book where his financial situation was carried by double entry. In the case of a worker, for example, this gave both his receipts in wages and what he owed for food, clothing, and supplies which he had received from the Company or from local merchants on the Company's credit. A fair portion of one such account, that of Francis Perry, the works carpenter, has survived. It shows debits for food, etc., as of given dates between May and September, 1653, of £148.9s.¾d. and credits by earnings of £137.8s.6d.[46] There were more than seventy-five of these individual accounts for full- and part-time workers. In all likelihood, similar accounts were kept for each of the Boston, Lynn, and Salem merchants and tradesmen who supplied the ironworks and who were paid, now in cash, now in iron, now in bills of exchange.[47]

Besides these financial records, which must have kept Gifford and his clerk well occupied, the agent was also responsible for production records. We know how the Undertakers wanted these kept, as shown by a number of samples or "forms of weekly account" which were offered in evidence in court when Gifford got into trouble. These cover a year's work in timber cutting and coaling and a week's activity at furnace and forge. The former was highly detailed and, if followed, would have given the Company a very complete and accurate picture of the appropriate areas of operations. The latter were simpler. They called, however, for a breakdown of costs, a record of sales, transfers, and weight losses incurred in processing, and data on the return on sales.[48] It is more than doubtful if Gifford actually kept records as good as these. He seems, however, to judge from a few actual specimens, to have come through with at least a reasonable facsimile. In his slitting-mill production record, for example, he carried bar iron turned over to the mill, month by month during 1652 but including

[45] Ironworks Papers, 42–46.
[46] *Ibid.*, 150–55.
[47] *E.g., Essex R. & F.,* IV, 437–38.
[48] Suffolk R. & F., 225.

an omnibus item of "Ironn deliuerred at Sundrey times," as debits. Against this he set the quantities of rod iron made and delivered, listed by date, destination, and weight, as credits. The latter, plus an item listed as "Losse in Slitteing the Ironn," exactly matched the former.[49]

Surviving sales accounting data are, as suggested earlier, thin in the extreme. So are those covering the intra-Company transactions between Gifford and Awbrey. The Undertakers had intended that each should have the right to inspect the other's books. There was also provision, apparently, for the annual exchange of summaries or abstracts. Awbrey, at least, got one such abstract from Gifford covering transactions in the year 1651.[50] It showed receipts by the agent of £2140.17s.6d. in money and bills and deliveries by him to the factor of 65 T. 7 cw. 23 lbs. in bar iron, 12 T. 1 cw. 2 qr. of rod iron, and 7 T. 4 cw. 1 qr. 25 lbs. of pots and cast ware. Provisions supplied Gifford by Awbrey were not included. Of these the latter said that he could not render an accounting "till wee haue examined and Cleered with the shops." One would give much for samplings from Awbrey's books to match those of Gifford's which have come down to us. One might gather from the lapse just mentioned that however soundly they may have been set up, they had been kept at least a little sloppily. Gifford's bookkeeping clearly got careless as time went on. Awbrey's was apparently less than good from the very beginning.

To this point we have been dealing with Gifford's regular business activities. There were, in addition, the peripheral but by no means unimportant responsibilities for community relations. Gifford was probably a Puritan by personal religious conviction. There is no record of his having had any brushes with the church at Lynn. During his term as agent he twice paid £10 as the Company's contribution to its support,[51] and his workers seem not so often to have been charged with failure to attend services as at other points in the ironworks' span of operations. By and large, the workers were now managing to keep out of trouble generally. Whether this reflected their more complete assimilation or the effectiveness of Gifford's leadership is, of course, open to conjecture. For all we know to the

49 Ironworks Papers, 67–68.
50 *Ibid.*, 26.
51 *Essex R. & F.*, II, 89.

contrary, at any rate, this ironworks agent seems to have been on good terms with his employees, his neighbors, and the Puritan magistrates until he fell victim to the mounting debts which were in time to prove both his and the Company's undoing.

Awbrey's work must also be regarded as part of the over-all ironworks management. His bookkeeping got off to a bad start. It presumably showed little subsequent improvement. Put in a hopeless position by the loss of all his papers, which were "burnt by the fall of a Candle in his warehouse or shedde," he was apparently unsuccessful in an attempt to build up at least a part of his books by copying from those of Gifford, in whose "faythfullness" he expressed complete confidence. According to one allegation, at least, he had eventually to admit, under questioning from John Becx and some of his partners, that he had received about £600 for which he could give no accounting and which he undertook to make up out of his own pocket as soon as he could.[52] Since at the same time he confessed that he had also drawn bills of exchange without authorization, it is less than clear why the Undertakers failed to follow through with their announced intention of discharging him. Certainly, he had not done the job of keeping the business side of the ironworks in the good order on which they had counted when they took him on.

What Awbrey did besides keeping what must appear to us as rather imperfect books is somewhat obscure. We have seen that he bought certain pieces of real estate in Boston. We know that he supplied the works, often going into debt as he did so,[53] although it is also clear that Gifford did some purchasing of supplies on his own. It was Awbrey who, with or without proper authority, drew bills of exchange on the Company which were taken up by local merchants in place of cash or iron. When, in the summer of 1652, the Company's local credit was in a slump, it was he who mortgaged the ironworks to guarantee payment on these bills of exchange.[54] He seems also to have played a key part in the sale of the services of the Scots servants who did not stay on as ironworks employees. The infrequency of his appearance in surviving court records should not lead to a conclusion that he had not been busy. To handle shipments of goods from the

[52] Ironworks Papers, 226–27, 257.
[53] Ibid., 142; Essex R. & F., IV, 96–97.
[54] Suffolk Deeds, I, 227–28, 232.

Undertakers, supplies from local merchants, and the greater part of the ironworks' output, and this in the midst of a shaky credit situation, interference from his principals and their local commissioners, and clashes with Gifford, was no mean task. That he figures so little in the records of litigation here and in England may mean that he performed it tolerably well.

Of the ironworks business as a whole, there remains to consider only the vital, if somewhat complicated, question of scale, particularly in terms of total investment, total operating costs, and total production. A sound estimate of the capitalization of the Company of Undertakers is not easily achieved. According to one claim of the English investors, they had spent £15,000 in their ironmaking venture.[55] In Gifford's opinion, what got handed over to the Company's creditors in the autumn of 1653—the plants at Lynn and Braintree with their stock, goods, and servants—had a value of twelve or thirteen thousand pounds.[56] Since both versions appeared in petitions for redress of grievances, exaggeration may be suspected. Analysis of data from other, and hopefully less biased, quarters makes it plain, however, that Gifford and Awbrey were presiding over a quite costly industrial enterprise.

Let us first attempt to distinguish between actual plant value and the sums spent in keeping the ironworks going, between fixed and operating capital. In the 1650 inventory of stock and tools, these assets were valued at £4302.13s.2d. According to a 1653 inventory, taken by "neutrals" when the courts were parceling out the Company's property among its creditors, plants, stock, tools, etc., worth £3961.5s.6½d., had been awarded to those with the largest claims. At that point, Gifford held an attachment on other assets whose value is not indicated but which served as security for sums due him and the workmen in the amount of £1363.14s.5d. Unless Gifford's stewardship had been almost incredibly poor, the plants could hardly have declined in value between 1650 and 1653. He claimed, indeed, that he had left almost £2800 worth of stock and tools alone when he quit the ironworks management. Though we choose not to accept this figure, which is cited in an otherwise demonstrably inflated version of his tenure as agent, a round-number estimate of from £4500 to

---

[55] *C. S. P. Col.*, America and West Indies, 1661–68, p. 17.
[56] Mass. Arch., 59/53.

£5000 for the value of the two plants and their equipment appears to be minimal.

In Gifford's statement of his business activities, which must be suspect because it was obviously offered to justify a record which had been found wholly bad by both the Company's commissioners and the Massachusetts courts, he claimed to have spent £7547.10s.11d. in money, £2115.15s.6d. in stock, or £9663.6s.5d. in all. Against this he set £8816.16s.6d. as the value of the "effects raysed and delivred," total production, in other words, plus various sums still due the Company. Adding this to his £2792.11d. of stock on hand, and to the value of assorted items which he felt ought fairly to count as assets, he came out with more than £13,000 on the credit side of his ledger. How much inflation is present in all this, we cannot tell with certainty. With respect to total costs, the area of our present concern, the surviving portions of his actual accounts seem to indicate expenditures of somewhat less than £4500 between January, 1651, and September, 1652, and of about £1800 during the first seven or eight months of 1653.[57] While here there are *lacunae* and items whose interpretation is difficult, these figures ought to be at least somewhat more reliable than those cited earlier. Making arbitrary allowance for the blanks and doubtful items and, again, guessing at a tenable round figure, one might set £7000 as a fair approximation to Gifford's expenditures of operating capital.

If we combine the latter figure with our estimated plant valuation, we get a total of £11,500–£12,000. This would presumably cover both Gifford's direct outlays and whatever portion of earlier expenditures was still being represented in the value of the ironworks plants. It is well short of the £15,000 which the Undertakers claimed to have spent. It is by no means inconceivable, however, that the outlays by Winthrop and Leader could have made up much of the difference. We do not know how much they spent. We know that expenditures in one area, the construction of Braintree Furnace, had been pretty much down the drain and would not be reflected in any of the assets figures cited above. Whether we accept the Undertakers' version of their total investment or the more conservative total which we have been working up synthetically from stray data, however, the sums

[57] *Essex R. & F.,* I, 294–95; VIII, 200–203; Ironworks Papers, 43, 139, 149, 250, 252.

involved are most impressive, given the time and place in which the investment was made.

What was the return on this investment? The question comes more readily than its answer. With production figures, as with fixed investment and costs, we must steer a course midway between wildly boosted claims and what we can infer from fragmentary but presumably more neutral estimates. The highest version of furnace production we owe to Gifford. Appealing for government support of further exploitation of New England mineral resources, and so anxious to prove that he was a well-qualified expert as hardly to let a few facts stand in his way, he announced that he had made and cast seven hundred tons of iron and ironware in a single year.[58] In no surviving ironworks data is there even an approximation of substantiation for such a figure. From Gifford we also get the next-highest estimate, this time in the 1655 document justifying his stewardship, from which we have already cited some data. Here he gave the following totals for production in all categories during what appears to have been his whole term as agent: 284 T. 18 cw. 1 qr. 20 lbs. of bar iron worth £5936.1s.6d.; 33 T. 8 cw. 2 qr. of rod iron valued at £935.18s.; 40 T. 13 cw. 4 qr. 21 lbs. of pots rated at £1203.6s.; and small ware, cast ware, and iron weights, quantities unspecified but worth £141.17s.3d., £93.13s.8d., and £47, respectively. The total value of all this, the "effects raysed and delivred" mentioned earlier, is given at £8357.16s.5d.[59]

Though we have already suggested that the document in which these figures appear is subject to suspicion, it is our only source of information on *total* production during the Gifford period. Its contents can be checked, at least in part, however, by setting against them the data available, in the appropriate categories, in the surviving portions of the Gifford accounts. Somewhat strangely, they stand up tolerably well when we do so. In certain details there is complete correspondence. At least a rough corroboration of the totals is obtainable by averaging out into twelve-month periods, now Gifford's lump figures, now those carried as individual entries in the accounts proper.

Combining data from these several quarters, setting them in terms of annual production, and cross-checking as best we can, we obtain

[58] S. P. Col., XV, fol. 43 (Gay Transcripts [M. H. S.], I, 33).
[59] Ironworks Papers, 249.

the following estimate of the ironworks' output in the Gifford period: ninety-six tons of bar iron, twelve tons of rod iron, and from twenty to twenty-five tons of cast- and hollow ware.[60] Except for the last-named categories, production at the furnace is not given. Sows and pigs were not made for sale. Assuming a more or less standard weight loss of ⅓ between crude and wrought iron, however, it is possible to work backward from the bar-iron figure and reach a total of 144 tons of sow- and pig iron.[61] This, plus the products cast in molds, is obviously far short of Gifford's claim of seven hundred tons! When we multiply our total quantities by prevailing prices—£20 per ton of bar iron, £28 for rod iron, and £25 for cast ironware—the cash value of a year's production works out to be £1680 for the bar iron,[62] £336 for the rods, and £500–£625 for the hollow- and cast-ware, or £2516–£2641 in total.

We have fairly decent data on the operating costs of the plant units involved, the furnace, forges, and slitting mill, between January 5, 1651, and September 29, 1652.[63] The supplying of these units with raw materials and fuel, and their operation, ran to £2322.17s. in the indicated period. This was by no means the whole story, however, since Gifford was also running a farm, keeping Scots and English indentured servants, maintaining watercourses, and the like. Costs in these areas amounted to £2165.15s.4½d. The total cost of operation of the ironworks was thus a matter of £4488.12s.4½d. If we bring this down to a twelve-month interval by averaging, it would appear as if Gifford had spent £2564.18s.6d. to make about £2600 worth of iron a year.

On this basis Hammersmith and Braintree ironworks had hardly

[60] The figures cited have been developed by arithmetical operations applied to the production and sales data contained in the Ironworks Papers, and checked, wherever possible, against such fragments and summaries as turned up in evidence in the complicated lawsuits to be discussed below.

[61] The weight loss obviously varies according to the quality of both pig and bar iron. No documentary evidence on the ratio prevailing at Hammersmith has been found.

[62] The rate cited for bar iron is that of the Court-designated "ceiling price." The rod-iron figure cited was derived by dividing the value of given sales by given weights of rod iron appearing in the Gifford accounts. The cast-ware rate is an average of an actual £30 for pots and an actual £20 for iron weights and other solid castings. The twelve tons of bar iron that got slit into rods have here been subtracted from the bar-iron total, lest they be counted twice.

[63] Ironworks Papers, 42–43.

been profitable enterprises. The record of assorted trials and tribulations which came to a climax in the sequestration of the Company's assets in the autumn of 1653—and the jailing of John Gifford—leads to the same conclusion. Only a blowing up of assets, and a recurring emphasis on potentialities rather than actualities on the part of the Company's third agent, could make the New England ironworks of the Company of Undertakers look like a financially successful business enterprise. The story of the troubles and an analysis of what lay back of them is best postponed until we have looked at other facets of the ironmaking venture, the plants themselves and the men who worked them. Here we have seen what the management of seventeenth-century colonial ironworks involved. A subsequent chapter will provide some grounds for assessing its calibre.

9

Hammersmith—the Technology

THE MAKING OF IRON in the seventeenth century was a quite sophisticated operation. The techniques of searching out and testing ores, of ridding them of their impurities, and bringing the metal to the appropriate composition, size, and shape were the product of generations of trial-and-error effort. Those employed at Hammersmith, and at its Braintree branch plant, were paralleled at many locations in the homeland. The Massachusetts exceptions to commonplace practice were few in number, and stemmed almost entirely from the necessity to adapt old and familiar processes to a new environment. The most striking element in the whole effort of the Company of Undertakers from the point of view of the historian of technology was the scale of enterprise, not its novelty. These men intended to set up as good a plant as currently available practice permitted. Hammersmith was conceived in the grand manner.

The effort, and the achievement, are the more impressive when seen in the prevailing climate of scientific and engineering activities. This was the borderland between alchemy and science, as even a skimming of the letters of John Winthrop the Younger will quickly demonstrate. Detailed information on the mineral surveying and ore testing carried out by this "scientist" is lacking. From old metallurgical treatises, however, it is possible to develop a synthetic picture of the way in which he and his workmen doubtless went about their tasks. As long as Hammersmith and Braintree Works operated, other men scoured the countryside in search of ore, tracking down rumors, making on-the-spot tests, and carrying likely looking specimens back to the plant for more systematic analysis and practical trial.

Surface appearances—rust on a rock mass, or rusty scum on a stagnant pool—would have been regarded as good omens, promising enough to justify testing of the rust-bearing or rust-producing materials for color, luster, hardness, and cleavage. To these, a Winthrop would probably have added streak, the color of the dust left on an unglazed tile in our day, on a touchstone, a piece of black marble, or other black stone in his. The yellow-brown streak of bog ore or limonite would have been a particularly useful test of the principal ore of early New England ironmaking. In all likelihood, too, a man of Winthrop's knowledge and experience would have known that bog ore was of middling specific gravity, or "intensive gravity," as it was called prior to the 1660's.

Above all, of course, these early assayers of iron ores were interested in their quality, in their ratio of iron to impurities. In this the principal tool was the loadstone. The pull of rock ores in their natural state might have served as crude index to relative richness in iron. For a sharper test, however, Winthrop and the better informed of his workmen and their successors probably reduced a small ore specimen to powder in a mortar, brought this powder to red heat, allowed it to cool, and then put it on a piece of paper. As the loadstone was brought close to the paper, the ratio between the iron particles attracted to it and the residue would have provided a somewhat better basis for estimating the worth of the ore at hand.

It was also important to learn how well the iron could be worked up. Here actual trial runs were made. When Winthrop's "finer" made iron from the Braintree bog ores during the preliminary location surveying, he probably used a small "furnace" made of stone, melting down ore and some form of fluxing material in a laboratory-size version of a bloomery forge. The product was a small quantity of wrought iron which could have been worked under a hand hammer, and broken for examination of its structure as it appeared in fracture. If it was neither brittle nor coarse grained, it would have been pronounced good.[1]

It was with techniques as crude as these that the pioneers of

Braintree and Hammersmith sought to test a form of iron ore that occurs in nature in considerable variety. There is a range in terms of physical mass from earthy particles to solid chunks. The useful content, a hydrated sesquioxide of iron $(FeO(OH).nH_2O)$, is on the order of from 35 per cent to 55 per cent in the specimens uncovered in archeological work at the Hammersmith site. There is also diversity in the conditions in which the ore that fed the furnaces at Braintree and Lynn was found. Some ore was dug up in what had come to be dry land. Some was taken from swamps, or scooped up with "floating shovels" from the beds of still bodies of water.[2] The latter is more correctly designated as "lake" or "pond" ore. Both types were processed at the Undertakers' plants, and while actual sources are almost impossible to pin down today, it is known that, over the years, the quest for ores was extended to far more distant sites than the nearby meadows of Adam Hawkes and other neighboring farmers.

To rid the bog ore of its impurities required smelting and the concomitant use of a fluxing agent. From ancient times limestone had been in common use to produce a more fusible slag and enable ironmasters to extract more iron from their ores. True, certain ores were so easily reduced as not to require a flux; in some cases, too, certain types of lava had been employed. And in the Forest of Dean, whence John Gifford and perhaps others of the Undertakers' employees had come, it was "cinder," the residue from old Roman bloomery operations, that got fed into furnaces along with the ore when it was not actually used as an ore *per se*. None of these materials was available in New England. While site excavations have produced some evidence of shells and coral, materials which were to be employed in many colonial American blast furnaces, it is all but certain that during most of its operating span Hammersmith Furnace used as fluxing agent not these but a dense igneous rock quarried on the nearby peninsula of Nahant.[3]

Here, one must assume, the absence of the generally known fluxes had posed a very real problem. Somewhere along the line someone had

[1] Johann Andreas Cramer, *Elementa artis docimasticae* . . . (Leyden, 1739), Eng. tr. *Elements of the Art of Assaying Metals* (London, 1741), 114, 136, 140–41, 337–39.

[2] Ironworks Papers, 192. The floating shovels were apparently purchased locally. Their use is not specifically indicated in the Company accounts.

[3] *Winthrop Papers*, V, 239. This appears to be the earliest reference to the ore that was to be carried in the Company accounts as "rock mine."

come upon a gabbro that appeared to have promise as an iron ore, and subjected it to trial, doubtless in the blast furnace itself. Since its iron content is well below 15 per cent, it could hardly have qualified as an iron ore, even as judged by seventeenth-century standards.[4] Presumably, however, it was sooner or later discovered that, when this material was combined with bog ore in the ratio of 40 per cent to 60 per cent, an iron of good quality could be produced without the use of any of the standard fluxes. Trial and error had led the men of Hammersmith to a new "right combination."

It took work to get out both the bog and rock ores. The bog ore was on or near the surface of the ground, and the bodies of water from whose bottoms the lake or pond ore got scooped up were probably not very deep. Getting out both types must have been hard labor, however. The Nahant gabbro, a dense and tough-grained rock, was doubtless "quarried" by drilling and the application of heat. Workmen, we know, were paid for carrying tools to Nahant "to breacke the ore," and for cordwood "to burne the ore at Nahaunte," and it is highly doubtful if the latter item covered an ore-roasting operation.[5]

The advantages of ore roasting were being debated well into the nineteenth century, and the whole subject is sufficiently obscure to suggest extreme caution in discussing ore roasting at Hammersmith. The Nahant ore was doubtless roasted before it was fed into the furnace. The physical qualities of the ore as quarried, the appearances of many specimens taken at the Saugus River site, and the documented presence of a "furnace to burne mine" make it plain that ore roasting was carried on at Hammersmith. It is not clear that the bog ore was roasted, although it was without question broken into pieces of workable size by hammering before it was fed into the furnace.[6]

The roasting, whether of bog or the Nahant "rock" ore, or both, presumably took place on a shoulder of the bank against which rose the blast furnace, and quite close to that unit into which the ore was eventually to be fed. The roasting "furnace," and its functions, prob-

---

[4] The Nahant gabbro, referred to by some geologists as a pyroxenite, comes in a low and high magnesia variety. An analysis of the latter by Dr. Fred Matson of Pennsylvania State University procured through the kind offices of Dr. Hobart M. Kraner of Bethlehem Steel Company works out to a total Fe by weight of only 9.88 per cent.

[5] Ironworks Papers, 89.

[6] *Essex R. & F.*, I, 295; VIII, 201.

ably had much in common with those of the Forest of Dean that Henry Powle described for the Royal Society in the following terms:

After they have provided their Ore, their first work is to Calcine it: which is done in Kilns, much after the fashion of our ordinary Lime-Kilns. These they fill up to the top with Coal and Ore, *stratum super stratum,* until it be full; and so putting Fire to the bottom, they let it burn till the Coal be wasted, and then renew the Kilns with fresh Ore and Coal in the same manner as before. This is done without Fusion of the Metal, and serves to consume the more drossy parts of the Ore, and to make it friable; supplying the Beating and Washing, which are used to other Metals.[7]

Neither archeological nor documentary data shed much light on this particular unit of the Hammersmith plants, however.[8] A glance at a specimen of the Nahant ore is enough to indicate why it could benefit from the roasting. If the bog ore also got roasted or "calcined," and the two terms are interchangeable, it was probably to extract earthy impurities which, if we may believe Swedenborg, could not properly be handled by the smelting furnace alone.[9]

The fuel was charcoal, the black porous form of carbon obtained by charring wood. To produce the enormous quantities involved took a special technique, rare skills, and careful and long-continued activity. The charring took place, ideally from May to October, in "pits" or carefully laid up mounds of wood that had been cut and corded on the Undertakers' large land holdings. Their erection was complicated. Radiating from a center "chimney" were four main tiers or layers, the "foot," "waist," "shoulders," and "head," formed, in the case of the first pair, of heavier lengths of wood, or "billets," set in place vertically or in vertical slope, and, in the case of the latter pair, of lighter pieces, the "lap wood," usually laid on horizontally. All of the stacking was handled with the greatest care, with chinks and crevices being plugged with smaller pieces of wood to achieve good

[7] Royal Society of London, *Philosophical Transactions,* XII (1677–78), 932–33.

[8] *Essex R. & F.,* I, 295. The base of a small stone structure, with what have been taken for a cast plate and a sow still intact, was uncovered at a point close to the charging bridge of Hammersmith Furnace. Determination of the height and over-all structure of the unit proved impossible, and it has not been reconstructed.

[9] Courtivron et Bouchu, *Art des Forges et Fourneaux à Fer, Quatrième Section.* "Traité du Fer, par M. Swedemborg (*sic*); traduit du Latin par M. Bouchu" (n.p., 1762), 68.

tight mounds ranging from ten to fourteen feet in height and from thirty to forty feet in diameter, and containing up to thirty cords of wood. To them, finally, was applied a covering of leaves or turf and a coating of "dust," the residue of former charring activities raked off as the basic site was being cleared.

The pile duly set up, the collier mounted it by means of a crude ladder, removed the center stake around which the chimney had been formed, filled the opening with wood chips, ignited them, and, after setting in place a cover or chimney stopper, took off for what was to be the last good night's sleep for some time. If all went well, the pit was safe for the first twelve hours. Thereafter, however, it had to be tended night and day, with the skilled craftsman snatching such sleep as he could in the nearby cabin in which he lived.

The collier's enemy was fire, the "live" fire which would consume and undo all the efforts to secure the "dead" or charring fire. Flames might break through the covering and destroy the whole pile. A gas explosion might blow off the cover. Strong winds might do the same, or cause uneven charring, however carefully selected had been the site, however well protected with "hurdles," or wind screens, the pit itself. Soft spots could also easily develop. These the collier had to find by the rather dangerous expedient of jumping up and down on the pile, and then reinforce them by readjusting the wood layers or replacing really bad areas with new wood, leaves, and dust. When charring was not proceeding evenly, the collier had to discover the weak areas with a long stake, or probing rod, and correct the difficulty by cutting draft holes. Too much draft, on the other hand, meant too rapid burning. Keeping everything in balance, achieving the blue smoke and the even settling of the pile that indicated that all was going well, took experience and almost constant attention.

Charring took five or six days, cooling, two or three more, and the latter required almost as great pains as the former. The outer covering had to be removed, a small area at a time, lest what appeared to be black and cold charcoal get rekindled. When, finally, the collier's trained eye told him that the charring was complete, the charcoal could be removed, or "drawn." Since even here there was danger of fire, the material was removed in small quantities, the accumulations being kept separate and at some distance from the main pile. No more was drawn than could be handled, wagon load by wagon load, and the pit itself, or what remained of it, was carefully watched throughout the proceedings. Eventually, of course, the charcoal was hauled off to the works, to the "coal house" located close to furnace and forge at Hammersmith.

Bog and rock ore, and charcoal—these were the raw materials of the Saugus River plant. All were fed into the furnace, the first of the two complementary units of the indirect process of iron manufacture. Its main function was to provide crude iron, which was converted in the second unit, the forge, to wrought iron. But from the molten mass confined in the furnace crucible also came hollow- and cast-ware items, such as pots, skillets, weights, and firebacks.

In terms of function, the furnace was a smelting and reducing unit, a stone stack in which iron ore was melted down, along with the fluxing agent, to separate the metal, iron, from the impurities with which in the ore form it was mixed, to strip it of its oxides. As the melting proceeded, the iron, being heavier, accumulated in the bottom of the crucible, and was run off through a tap hole into sand trenches where it cooled in the shape of sows or pigs, or was carried out in ladles and poured into clay molds.[10] The impurities, being lighter, floated like scum on top of the fiery bath in the crucible, and were raked off as slag.

In terms of construction, the furnace was a structure about twenty-six feet square at the base, twenty-one feet high, and with outer walls sloping inward as they rose. Made of granite and other local stone bonded with a clay mortar, it rested on level ground into which a subterranean drainage system had been cut to guard against the dampness to which the water that drove its big bellows wheel made it peculiarly susceptible.[11]

Evidence of the seriousness with which seventeenth-century ironmasters took the threat of moisture is available in several quarters. We saw earlier, for example, the emphasis that Sir Charles Coote, the proprietor of vast iron enterprises in Ireland, had placed on this

[10] The sand beds into which sows and pigs were run and the area in which the clay molds were buried were found nicely defined in excavations at the Hammersmith furnace site.

[11] Dimensions and other construction details derive from the archeologist's findings.

aspect of furnace building as Winthrop was about to set up his own works.[12] Plot, in his *Natural History of Stafford-shire*, was even more emphatic, saying:

Tis also of importance in melting of *Iron Ore*, that there be five or six *soughs* made under the Furnace . . . to drain away the *moisture* from the *furnace*, for should the least drop of water come into the *Metall*, it would blow up the *furnace*, and the *Metall* would fly about the Workmens ears; . . .[13]

In the interior was a kind of stack or chimney, roughly egg shaped, six feet in diameter at its widest point, the top of the "boshes," or downward sloping surfaces, which supported the charge of ore, flux, and fuel.[14] Below the boshes was a square crucible, lined, like the cavity in which the burning occurred, with good refractory sandstone.[15] Between the lining and the outer masonry ran a pair of inner walls, one extending to full furnace height, the other only as far as the boshes, which it helped to support. The spaces between these walls were filled with sand, clay, and rubble as a cushion for the expansion and contraction of heating and cooling. Two of the outer walls had large and deep arches. Through the smaller one passed the noses of the eighteen-foot bellows and the "tuyere," or conduit, through which they delivered their blast to the crucible. Under the larger arch was the vital working area of hearth and casting floor.

The crucible, the reservoir for the molten iron, was eighteen inches square at the base but broadened out to twenty-one inches as it reached its full height of three and one-half feet. A projection of its lower portion, the "forehearth," consisted of two side walls and a forestone or "dam." Above, and set back from the dam, was a stone curtain wall, or "tymp," whose bottom edge came down lower than the top of the dam. Through the opening thus formed, a workman ladled

---

[12] See p. 51 above.

[13] Robert Plot, *The Natural History of Stafford-shire* (Oxford, 1686), 162–63.

[14] No archeological evidence for the shape of the furnace-stack interior was found at Saugus. In ascribing to it the round, rather than the square, I am following what I take to be the majority opinion of English and American scholars of the history of iron technology.

[15] To judge from the record of difficulties in finding hearthstones for New Haven Furnace, this sandstone was doubtless imported. The samples found in great quantities at the site bear no resemblance to materials to be found anywhere in New England, so far as I am aware.

off the iron for mold casting and, with an iron rod, or "ringer," pried away slag that stuck to the crucible sides and corners or accumulated around the tuyere vent. For protection against the wear and tear of such operations, both the tymp and the dam were sheathed with iron plates. Slag removal was merely a matter of raking the material over the dam stone. To tap the iron, however, required the breaking out of a clay plug inserted in a narrow space left between one of the forehearth side walls and one end of the dam.

Besides all this reasonably complicated masonry, the erection of the blast furnace involved work in timber and in leather. Between furnace top and the bluff beside which the furnace stood, ran a heavy timber bridge, over which the charging materials were carried. On three sides of the top was a wooden wind screen, set up to provide at least some shelter for the workmen who had to pour basket after basket of ore and flux and fuel into a charge hole that belched smoke, sparks, and, at times, even flames. Wrapping the lower portion of the furnace was a crudely framed "casting house," which provided at least some cover for the trench and mold casting, and perhaps for the making of the molds as well.[16] It also housed the bellows and their frames, those eighteen-foot leather lungs which alternately inhaled and exhaled the air without which the high temperatures required in the smelting would have been impossible. Their deflation was accomplished by means of cams on the furnace water-wheel shaft, their inflation, by counterweights, stout wooden boxes filled with stones and mounted on moving beams that extended beyond the casting-house roof through holes cut to accommodate them.

To the men who worked the furnace, it was an "ornery" beast that had to be attended and "coaxed" day and night, and for long stretches at a time. When first put in operation, the furnace was seasoned by starting a fire in the crucible, this to make sure that all was perfectly dry. Two or three days later the first loads of real fuel were brought on and the furnace began to "blow." Into the giant cavity, on the order of 340 cubic feet, went the charge, or burden, layer on layer.[17]

[16] While references to "the potter's house," etc. were found for certain of the latter New England ironworks, neither archeological nor documentary data on the facilities for mold making at Hammersmith have been discovered.

[17] James Pilkington, *A View of the Present State of Derbyshire: with an Account of Its Most Remarkable Antiquities* (2 vols., Derby, 1789), I, 134–35. The order was, working from the bottom, fuel, ore, and flux.

Approximately three tons of bog ore and about 265 bushels of charcoal were required to make a ton of iron. The prevailing ratio of ore to flux seems to have been on the order of 60 per cent to 40 per cent.[18] Since, once in operation, work continued without let up for as long as thirty or forty weeks, and new increments of ore, fuel, and flux were called for at intervals determined by measuring the settling down of the charge with a gauge inserted at the charge hole, the "fillers" must have been very busy men indeed.

Down at the "business end" of the squat masonry pile, there was also plenty to do. There was the blast to keep well adjusted, a matter of regulating the flow of water over the wheel by means of a wooden lever connected to the sluice gate. There was the clearing out of slag, the preparation of the casting trenches, the careful placing of the clay molds, buried in earth except for their sprues and risers.[19] Once or twice in each twenty-four-hour period came the crucial event for which all else was preparation: the slag skim was removed; the clay plug was broken; and out gushed the stream of molten metal. Running into the V-shaped ditch that had been furrowed with a triangular hoe called a "ship," it stewed and sputtered and slowly cooled and hardened. Back in the furnace, however, a new "batch" was already in process. The making of iron was not really finished until at last the furnace "blew out." Down to that point when, normally, crucible and part of the lining had to be replaced and the whole series of tasks recommenced, there had been long days of hard work and many an interruption in the night's sleep, snatched, as best the workers could, on straw pallets in the casting house.

The waste, the slag, was carried off and dumped in the swampy area running down to Saugus River. The iron, both in the form of sows and pigs and of hollow and cast utensils, was weighed out with a great steelyard. Estimates of furnace production by time period appear elsewhere. Here we need point out only that it was, by present-day standards, modest in the extreme. The lower portion of the crucible and the forehearth had a combined capacity of about five

[18] It has not been possible to work out consumption of ore, flux, and fuel from surviving Company accounts. My estimate of the yield from the bog ore derives from the experience of later New England furnaces.

[19] Clay fragments and sprues and risers were found in vast quantities in the archeological operations. A selection of the more interesting specimens is on display at the Museum of the Saugus Ironworks Restoration.

cubic feet, sufficient, if all space were utilized and all slag removed, to hold about a ton of liquid iron. Surviving documentary data include one reference to a sow weighing about five hundred pounds and indicated as the product of twenty-four hours' furnace activity.[20] The largest one excavated at the Hammersmith site, however, is about fifty-two inches long annd weighs but two hundred ninety pounds.[21] The reasons for what appears to have been a falling short of theoretical maximum capacity operations are wholly obscure.

Work at the furnace reached its most dramatic heights at the times of tapping. Most of the activities carried on there were, at least to the layman's eye, largely a matter of tending. For the men of the seventeenth century, however, a working blast furnace held much the same fascination that a steel mill does for us today. We may readily understand why Gerard Boate, writing of ironworks in his *Ireland's Natural History,* published at London in 1652, could say of the flame rising from a blast furnace in the night that it "maketh a terrible shew to travellers who do not know what it is."[22] It is in no way strange that John Lucas recalled the story of Tubal Cain's seeing iron running from a burning mountain and being prompted to "invent" metal casting as he described the tapping of a furnace in Warton Parish. He spoke of the iron as:

A Torrent of liquid Fire; made so very fluid by the Violence of the Heat, that when it is let out of the Receiver or Hearth, by breaking a lump of Clay out of a Hole at the Bottom thereof, with a long Iron Poker, it not only runs to the utmost Distance of the Furrows, but stands boiling in them for a considerable Time.[23]

He saw beauty in the ironmaking operation, even in the cooling of

---

[20] *Winthrop Papers,* V, 246, where "sowe" has been incorrectly transcribed as "some."

[21] It is possible, of course, that this and other sows did not represent the full volume of a single tapping of the furnace. It was not uncommon to run small trenches off the main furrow, with the iron cooling in the former being called pigs, that in the latter, sows.

[22] Reprint in *A Collection of Tracts and Treatises Illustrative of the Natural History, Antiquities and the Political and Social State of Ireland,* I, 198.

[23] John Lucas, *History of Warton Parish* (1710–40) (J. R. Ford and J. A. Fuller–Maitland, eds., Kendall, 1931), 60. Most of the comment ought properly to be credited to Plot, *op. cit.,* 162.

the sows. No description of Hammersmith Furnace or of any other part of the Lynn works has survived. For many a Massachusetts Puritan, and many a Massachusetts Indian, however, the furnace scene must have exercised a potent fascination.

Hammersmith Forge, the second element in the Saugus River indirect-process ironmaking operation, was a most impressive plant. to judge from both archeological and documentary evidence. There were many variations in the operation by which the cast iron from the furnace got converted into wrought, as many as fifteen by one authority's count.[24] In essence, however, all of these can be brought under two general headings, depending on whether or not the job was carried through all its stages with the heat from a single hearth or with that of a pair of hearths worked in tandem. The former is called the "German," the latter, the "Walloon" process. It was the Walloon process which was standard in England in the seventeenth century. It was therefore this two-hearth system which was set up at Hammersmith, and at Braintree Forge as well. In these forges, there were both "fineries," hearths to melt down the pigs, decarburize the iron by oxidation, and provide the heat for bringing the wrought iron to a semifinished stage, and "chaferies," hearths in which the malleable iron got reheated for further welding and drawing out into finished bars. More accurately, Hammersmith Forge had two fineries and a chafery, and at Braintree Forge there was one finery and one chafery.[25]

Both types of hearth looked much like giant-sized blacksmith's forges. Both had a crucible formed of iron plates, each with its proper name, and all most carefully arranged, with leather bellows driven by water power, and a stack, or chimney. The base was of stone laid in clay. The chimney was of clay daubed over laths on a timber frame.[26] As structures, the finery and the chafery differed mainly in size, the

[24] P. Tunner, *Die Stabeisen-und Stahlbereitung in Frischherden* (2nd ed., 2 vols., Freiberg, 1858), as abstracted in an English translation in John Percy, *Metallurgy: . . . Iron and Steel* (London, 1864), 580–619.

[25] *Essex R. & F.*, I, 294, 295; VIII, 200, 202; Ironworks Papers, 35. Two giant anvil blocks were found by the archeological crew. Other pieces of archeological data, including the size of the forge building, watercourse outlines, and the assumed location of the fineries and chafery, plus the absence of documentary reference to a second hammer, made the actual use of two hammers highly questionable. The newly reconstructed forge therefore stands as a single-hammer unit.

[26] Ironworks Papers, 70, 115, 130.

latter being somewhat larger in crucible, and therefore in the bellows which provided the needed blast. Their more important point of difference, one of function, will shortly be apparent. Both hearths, too, were worked in conjunction with the hammer, a mighty machine consisting of an oak helve with a cast-iron head, mounted in a "hurst" held in great wooden legs, raised by cams on the water-wheel shaft and depressed by gravity and the bounce from a wooden spring which it struck at the height of its rise. Because of the enormous stresses involved, all of the carrying and reinforcing members, oak beams carefully mortised and wedged, were of huge dimensions. So were the iron anvil and the timber block, all but buried in the ground, mounted on interlocked oak beams, and tightly packed with clay, on which it rested.[27]

Work at the finery started with the piling of charcoal atop a layer of slag in the crucible. The fire lighted and the blast turned on, the pig was gradually brought up to the heat through a hole in the finery wall. Melting off at the end, the iron trickled down into the fiery bath, drop by drop. Now the finer's highly skilled eyes and hands were called on. With a long iron bar called a "ringer," he had to stir and work the fused metal, doing his best to achieve even melting and to keep the surfaces of the hearth plates and the tuyere mouth clean. Eventually, the metal settled in the bottom of the hearth where, in the lower temperatures there prevailing, it became a semisolid mass. Some of the residue of impurities, the forge slag, or cinder, was saved to provide a slag bath for further operations. If too much accumulated, however, the surplus was tapped off through a slag hole.

Next came the refining proper. This began with a second fusion, the semisolid mass being broken up, and those parts not yet sufficiently decarburized being again oxidized under the bellows blast. The results satisfying the finer, he went on to what was actually still another or third fusion. This time the metal melted down and formed pasty lumps in the bottom of the hearth. After about an hour's "cooking," these were gathered together and kneaded into a ball of iron called a "loop."[28]

[27] The block and its cross-timber base, found in an excellent state of preservation, are on display in the Restoration Museum.

[28] My account of the forge and its operation owes much to Dr. H. R. Schubert's superlatively fine paper, "Early Refining of Pig Iron in England," read before the Newcomen Society at the Iron and Steel Institute, London, January 9, 1952, and made available to me in typescript.

What followed was the operation called "shingling the loop." The ball was taken from the finery hearth with tongs and placed on a nearby iron plate on the forge floor. There it was beaten with a sledge hammer to consolidate it and remove surface charcoal and cinder. It was next dragged over an iron-plated path to the big helve hammer. Under its blows, at first slow and gentle, then faster and more powerful, the iron mass was further consolidated, its impurities being literally squeezed out in the form of cinder, and its shape being changed from rough round to thick square called a bloom. This was then cut in two and the pieces, the "half-blooms," were ready for further processing.[29]

The half-bloom now went back to the finery for what was primarily a welding heat, not another melting down. Even as it was getting its last finery heating, still another pig was being brought up to the crucible edge, ready to start its sequence of refining operations. After about an hour's heating the half-bloom was removed from the finery, taken to the power hammer, and forged into the shape of an "anchony," a mass vaguely resembling a more or less square-ended dumbbell. The middle portion was of the size and shape of the final product, the finished bar. The ends, however, were still rough, unfinished masses, one larger than the other.

The heating required for the finishing off of these ends was provided by the second hearth, the chafery. Here, too, the fuel was charcoal, burning under blast. First the smaller "head" was heated for about a quarter of an hour and then brought down to the desired shape under the big hammer. A similar treatment of the larger, or "mocket," head followed, although in this case two heats were required because of the greater mass of iron involved. The ends finished, the iron mass which had been carried from half-bloom to anchony emerged as a single bar, the form which was by all odds the principal sales item of the Saugus River plant.[30].

The refining of iron was obviously complicated; indeed, as late as 1775 it was described as "the most difficult operation in all metallurgy."[31] It was also dramatic in the extreme. Within a building

[29] *Essex R. & F.,* VIII, 201–202.

[30] To judge from fragmentary data, only about one-eighth of the wrought iron from the forge went through the rolling and slitting mill to emerge as nailer's rod.

[31] Quoted by Dr. Schubert in translation from J. A. Cramer, *Aufrangsgründe der Metallurgie* (2 vols., Blankenburg and Quedlinburg, 1775), II, 154.

roughly forty feet square could be heard the quick flapping of bellows, the thudding and ringing of the hammer as it rose and fell on the iron being processed. Outside, there was the whirring and splashing of four water wheels, two to a side in parallel courses. In the semi-darkness the fires in the hearths burned brightly, their changing colors both useful as index to the quality of the ongoing operation and beautiful. At one stage what looked like a fiery ball was being dragged over the floor from hearth to hammer and back. At others, sparks went flying as strong workmen hammered away on glowing masses. On occasion, as the hammer fell, a stream of liquid fire was projected from cavities in the iron high above the workmen's heads. Melting, hammering, reheating, forging—these activities were carried on by skilled craftsmen and their helpers in the midst of a great din, an almost endless fireworks display, and seemingly enormous confusion.

Just easterly of the forge at Hammersmith stood a slitting mill or, more accurately, a rolling and slitting mill. Of all the plant elements this is perhaps the most interesting historically. While the first machines of this general type were built about 1513, in the vicinity of Liége, their development and spread to other areas were slow. By one authority's estimate, previously cited, there were less than fifteen slitting mills on which data have survived in the world when one was set up on the banks of Saugus River.[32]

Its general function is clear from the first English patent, granted in 1588 for "a quicker and more apt and speedye waye . . . for in and about the cuttyng and makyng of yron into small barres or roddes to serve for the makyng of nayles and other thyngs. . . ."[33] The incorporation of the slitting mill in the plans for Hammersmith made perfect sense. The demand for nails in the building of the wooden houses and barns that became and remained standard in New England was enormous. The manufacture of nailer's rod by a drawing out of bar iron under the hammer was most expensive of time and labor. Product demand and industrial efficiency drove the men of Hammersmith to a machine which, while expensive, complicated, and difficult to keep in adjustment, might go a long way toward turning out the "raw material" from which farmers, working in off season

[32] Chapter I, n. 5.

[33] Rhys Jenkins, *Links in the History of Engineering and Technology: the Collected Papers of Rhys Jenkins* (Cambridge, 1936), 13.

by their firesides, and others, might make by hand the nails and spikes basic to frame building.

The work of the machine was to flatten and draw out bar iron and then to cut the product, the "flats," into strips or rods. The former was accomplished by passing hot iron through a pair of rollers, the latter, by running the flats between sets of opposing disk cutters. The most direct way to accomplish this was to erect a pair of water wheels in parallel courses, one on either side of the mill building, and get them running in opposite directions, and at the same rate of speed, in such a way that the shaft of one would turn the upper, the shaft of the other, the lower set of rollers and cutters, the rollers and cutters being set up side by side and joined. To judge from archeological evidence at the Hammersmith site, however, its slitting mill was powered by only one watercourse. Given the job that had to be accomplished, we must infer that in its case topographical conditions had dictated the use of indirect power transmission.

To men who had built gristmills, indirect power transmission was nothing new. To rig a machine which would handle the stresses of working tough iron, however, must have called for no small degree of technical competence. In a single wheel pit two huge water wheels had to be set up. One wheel delivered its power directly, providing the power for one set of rollers and cutters. The other, however, turned a giant cogwheel geared to a lantern pinion of equally impressive proportions. The lantern turned a shaft which was mounted over and parallel to the shaft of the first water wheel, and delivered the energy for the other set of rollers and cutters. As in the simpler or directly powered type, the working elements, the pair of rollers and the pair of slitters, stood side by side in frames, their axles joined by couplings. Finally, as though this were not complicated enough, there was also a great shears for cutting iron into appropriate lengths, which was operated by a cam on one of the water-wheel shafts.

Within a crude frame building, then, was some highly sophisticated machinery. It housed, in addition, a stone "furnace," in essence a crude, wood-fired, reverberatory furnace, in which the bar iron destined for rolling and slitting was heated, and, for the opposite end of the operation, a stout table on which the finished rods were bundled, and a simple little hearth for heating the iron straps with which the bundles were bound. The bar iron from the forge was cut cold in the shears into two- or three-foot lengths, and the pieces stacked in the furnace and heated for as much as four hours. These were then drawn out between the rollers, probably in two or more passes, to a length of eight or ten feet. In the process, there was an increase in width and a decrease in thickness, perhaps on the order of from three to five and from one and one-half to one-fourth inches, respectively. The flats were passed back over the frames by the workman who took them as they came through the rollers, and then sent through the slitting unit, where they were cut lengthwise by from three to six steel-edged forged disks, depending on the rod dimension desired. Changes in the thickness of flats and in the size of finished rods could be made only by dismounting the equipment and inserting rollers of different diameter, and a new brace of disks.[34] A stream of running water tapped from the water-wheel flumes trickled down to both rollers and cutters as the iron was being run through them. At a lower level, again to depend on evidence found at the site, a workman cleaned and straightened the rods and put them up in bundles weighing about sixty pounds, ready for sale.

Maintenance of machinery worked with the close tolerances of the rollers and slitters was no easy matter. The rotation speed was governed only by control of the flow of water on the big wheels. Keeping their shafts in alignment and getting cogwheel and lantern to mesh properly must have demanded frequent adjustment of the plummer blocks on which their axles rested. Roller surfaces had to be kept smooth and cutting-disk edges sharp, and here both a blacksmith's forging skills and backbreaking work with a grindstone were called for. With little question, too, breakdowns were probably commonplace in what was obviously a heavy power operation.

Furnace, forge, and slitting mill—these were the main plant elements of Hammersmith. To describe them and give a general picture of what went on in them, unit by unit, has been necessary. It is also important, however, to see the plant complex as a whole. Archeological evidence has established, almost beyond question, that Hammersmith was conceived as an integrated whole. It had unity as an engineering construct. This is attested by a layout which permitted the flow of materials from the ore piles at the furnace head to the finished bars and rods stacked in the warehouse ready for shipment that calls

[34] Plot, *op. cit.*, 163.

181

to mind a modern assembly line. It is also indicated in the articulated water-power system which, down to this point, we have been considering only in passing and in terms of its component parts.

When Richard Leader first came to the area on which Hammersmith was to rise, he apparently found a point where a bank of glacial origin shaped vaguely like an amphitheater rose above and somewhat back from the swampy banks of Saugus River.[35] Though it would take clearing and filling in some portions, the "amphitheater" offered a means of setting up the three main units in such a way that all would have adequate water power and yet permit materials to be moved from furnace through to forge and slitting mill on level or downward-sloping surfaces. Movement of cast- and hollow-ware from the furnace, of bar iron from the forge, and of rod iron from the slitting mill to the warehouse was, again, down. The marshy shore line offered a fine place for the disposal of slag, a problem as real for this seventeenth-century ironworks setup in a comparative wilderness as it is in many steel mills of our own day. The over-all immediate working area was a bit crowded. There was at least the potential danger of the slag dump's crowding the river boat-docking facilities. This could be cured, as it was in time, by digging out a kind of boat basin. There was the possibility of the washing away of the shore line, but a reinforcing with a timber "wall" could provide for this, even as corduroy paths would facilitate travel over adjacent soggy areas.[36]

The water-power system posed problems, too, but work, much work, would enable the men who laid out Hammersmith to equip it with a water supply dependable in all but periods of extreme drought. The river had first to be dammed. At a point well upstream of the works site tons of earth were brought in to form a dam, well over one hundred feet in length, eighteen feet in height, and about seventy-five feet in thickness at the base. On the water side this earthen rampart was lined with stone from top to bottom.[37] The river water backed up to form a pond that extended over many acres. A sixteen-hundred-foot canal, still visible in places, ran from the pond to a

[35] This follows the interpretation of the original ground plan worked out by Roland Wells Robbins, the Restoration archeologist.

[36] Archeological evidence is quite convincing. Both the extent and the workmanship of the timber reinforcement were impressive.

[37] Ironworks Papers, 35; Essex R. & F., XXXIX, 97–1.

stand-by reservoir close by the plant. From the reservoir, in turn, extended a watercourse that powered the furnace wheel, two that supplied the forge, and another the slitting-mill wheels. In all cases, presumably, the water ran from reservoir to wheels in raised wooden flumes. After it had done its work, it flowed down to the river in open channels, bridged at a number of places for movement from one to another of the plant units. The laying out of this large system must have been, by seventeenth-century standards, a major engineering undertaking.

No documentary data on wheel construction have survived. In recent excavations, however, a fair portion of the furnace wheel and essentially all of the pit in which it turned were found intact. It is therefore certain that the furnace bellows had been driven by a six-spoked overshot wheel between sixteen and seventeen feet in diameter and about two feet wide.[38] The craftsmanship of some colonial wheelwright is abundantly plain in the excavated specimen. The dimensions and type of the other wheels are not definitely known, although it is clear, both from general archeological evidence and from their known or assumed functions, that all were quite large, that one was an undershot, the others overshot or pitch-back. All must have been as well constructed as the furnace wheel.

Much of the work at Hammersmith took no power beyond that furnished by men's muscles. At a point in the vicinity of the warehouse stood a blacksmith shop, in which repair and maintenance work on tools and machinery was carried on at a forge presumably equipped with a hand- or foot-driven blast.[39] In the same area, too, was a saw pit, at which a workman labored to produce the boards needed for upkeep and additional building.[40] If the turning and firing of the clay molds was not carried on in the blast-furnace casting shed, there must have been, somewhere on the premises, a shop where the potter toiled, and again, in all likelihood, without benefit of water power.

In the total ongoing ironmaking operation, there were numerous other minor activities. The bar iron from the forge and the rod iron

---

[38] A good portion of the original wheel, found in an excellent state of preservation, is now on display at the Restoration Museum.

[39] *Essex R. & F.,* I, 295.

[40] Essex R. & F., III, 51–2.

from the slitting mill, for example, were weighed, both to determine weight loss between processes and to record production and sales and transfers. There is archeological evidence, too, of activities whose nature cannot be ascertained either from the actual findings or from documentary evidence.[41] And finally, of course, there was, downstream of the furnace, the hammer and plating mill, or toolmaking forge, of Joseph Jenks, here regarded as no part of Hammersmith proper, however significant the work that master blacksmith there carried on.

Such, as best we can determine it, was the ironworks on the banks of Saugus River, such, the main operations in, and related to, the making of iron under the aegis of the Company of Undertakers and their successors. One would give much to be able to bring the canvas to life, to see smoke rising from the chimneys, or workmen moving heavy loads in wheelbarrows, to hear the din of pounding hammers and swishing bellows, or a Richard Leader bellowing orders to his employees. We cannot. At this point, however, having surveyed Hammersmith in terms of the tasks and trials of management, and of a high order of accomplishment in engineering and technology, it is time to turn to the workmen, skilled and otherwise, who made the iron—and trouble for their supervisors and their neighbors.

[41] These included small piles of stone bearing evidence of exposure to great heat but not recognizable as hearths; portions of foundations, with a massive rectangular cast-iron plate forming the floor for one set that was not far from the furnace watercourse and close to the base of the bank against which the furnace had been erected; etc.

10

Hammersmith—the Workers

THE IRONWORKS STORY as it has been handled to this point has had the promoters, the capitalists, the Massachusetts public officials, and the agents or managers in the foreground. In a major sense, however, Hammersmith and Braintree Works were the products of the brains and brawn of the workmen who staffed them. History had not been kind to these men. They came here, worked, raised families, died or moved on to other ironworks or to other fields of endeavor, all in comparative obscurity. In the recording of their activities which has come down to us, it was, in the great majority of cases, the negative that got accentuated. We now know, for example, vastly more of their bad actions, their brushes with the law, their clashes with the Puritans with whom Fate had brought them to live, than of the positive accomplishments of lives spent in hard toil and painful adjustment to a new physical, social, and religious environment. We know more than a little of what their neighbors thought of them. We know almost nothing of their attitudes toward their betters and their peers.

Much of what we do know has been put to strange uses by some of the descendants of these ironworkers. The negative, and even the plain and homely, have been conveniently glossed over by people who, though American, have been reluctant to admit that their progenitors had been "mere" workers or indentured servants. A blacksmith for whom no one need apologize has been raised by his remote progeny to the dignity of master mechanic of Hammersmith. Members of another family, as conspicuous for their defiance of Puritan mores as for their great skill in wrought iron, have been made as respectable as many of their descendants managed to become. Wisps of data have been so inflated as to make the progenitor of this family or that into a major ironworks figure, or even a pillar of church and community.

Our purpose here is not to stress the negative, not to topple ancestors from the positions of pre-eminence to which misplaced filial piety has carried them, not to "debunk." It is to depict, as accurately as surviving data will permit, the working situation and the manner of life of a group of men who may fairly be regarded as prototypes of that mighty army of American industrial workers. What happened at Hammersmith and Braintree is, in many respects, Chapter I of a great story. Here were the problems of assimilation of "foreign" elements into a "settled" population. Here were, in several instances, the first steps in the transition from lowly status to riches and respectability. Here were industrial housing problems. Here were the development and passing on of skills. Here was geographical and vocational mobility. If we can analyze such phenomena, particularly in that small but nice case-specimen which was Hammersmith, we may help to shed light on a still somewhat neglected area of American history. We may be able to show that the joint contribution of obscure figures had been as important to the development of the ironworks and of Massachusetts as that of individual great figures with whom older historians tended almost exclusively to deal.

What kind of people worked at Hammersmith and Braintree? Immediately, we must put them into categories—skilled and unskilled, free and indentured, full- and part-time. The unskilled and part-time workers were, in the main, settled residents, ironworks neighbors, who felled trees and hauled timber, charcoal, and ore, or who plied trades which, though specialized, were not those involved in the making of iron, but such as carpentering, and tailoring, for example. Except for these men, to whom we shall shortly turn, we are dealing with craftsmen specially recruited and imported from the ironworking districts of England and Wales. Specific points of origin for even the more conspicuous have, with but few exceptions, defied the researches of the most determined of genealogists. Winthrop's original working staff is identifiable only by inference, and then only to a quite limited degree. We know even less of the background of later recruits to the ironworks labor force.

Certain generalizations are possible, however. Skilled iron craftsmen were in strong demand in the Old Country. Finding them had been a grim task for management. Those who could be uncovered and persuaded to emigrate were, with but few exceptions, young, willing to take risks under the influence of promise of higher wages and other long-range opportunities in a young settlement, and, most emphatically, *not* Puritans. The great majority came here with wives and children, or managed later to get them over, or married and raised families in New England. Bachelorhood, for ironworkers as for their neighbors, was rare. Family structure followed the general colonial pattern of many births and at least five or six surviving children in many instances. In some cases we find family teams—working brothers, or fathers and sons—although the data blanks on the actual transit from the Old to the New World make it impossible to tell whether they came here together or separately. Nearly all came originally in indentured, or contract-labor, status but managed to round out their terms of service, and became free skilled workers. Some carried with them, or soon managed to accumulate, small amounts of capital. Most of the workers, including even the most highly skilled, were illiterate. The total number of ironmaking specialists was unquestionably small.

In presently available data no less than 185 men can be identified as having worked for wages or under indenture in, or resided at, or been paid for services to, the ironworks at Lynn and Braintree over the whole span of their operations. Of these, only about one in five was a real ironworker. Counting employees at both plants, and including a few doubtful cases and clerks and managers who are not known to have had a financial stake in the ironworks, we find no more than thirty-five men who seem to qualify as full-time employees handling jobs directly related to the making of iron. All the rest were supernumeraries or nonspecialized indentured servants.[1] Owing to the

---

[1] On the basis of regular wage payments for skilled ironwork and residence in Company housing, I have classified men as full-time specialists. Conversely, receipt of wages on an occasional basis, normally for general services, and settled residence in

lacunae in the surviving accounts and to the by no means infrequent exchange of workers between Hammersmith and Braintree, not even a crude breakdown by year and plant is possible.

It is not easy, indeed, to distinguish even among vocational specializations. The neighboring farmers, in addition to the general chores they handled normally, occasionally took on jobs which usually fell to the regular ironworks employees. The latter, particularly in times of plant shutdown, often joined them in the forests, at the mine pits, and in the work of carting and hauling. Even in the activities connected with ironmaking proper, there was little specialization. Many of the workers were Jacks-of-all-trades, worthy sires of a long Yankee strain of specialists in versatility. This is not at all strange. It is an obvious consequence of the labor shortage inevitable in a frontier community. It does, however, pose problems when it comes to attempting a job analysis for the ironworks and the various units of which they were composed.

Of the thirty-five men who made up the effective working nucleus of New England's pioneer iron enterprises, two, judging from the wage accounts, were furnace specialists, Roger Tyler and Richard Post. Tyler was in all likelihood the founder, the man in charge, in Gifford's time. Deserving of a special place in our story as the worker who "layed the first stone of the foundation of the furnace at Hammersmith," his principal task was "blowing," for which he earned £54 at the rate of forty shilling a week during 1651. He was also paid for many odd jobs, such as basket making, felling timber, dressing the bellows, etc.[2] Post, who figures but little in the account fragments, was "blowing" at the furnace in 1657 or thereabouts, and was probably Tyler's successor, immediate or otherwise.[3] Thomas Beale and Thomas Wiggin kept the furnace filled, doubtless under Tyler's direction, but neither seems to have been a skilled ironworker.[4] John Diven, the potter, was certainly a specialized craftsman, but we know nothing as to how or when he was paid.[5] At least a half-dozen

---

Lynn or Braintree, especially when it antedated the establishment of the ironworks, were considered as indicative of part-time or supernumerary employment.

[2] C22/307/2, Deposition for Gifford; Ironworks Papers, 52, 62, 77–78, 83, 120, 140.

[3] Winthrop Papers, 13.73.

[4] Ironworks Papers, 62–64.

[5] Essex Deeds, VI, 111; Ironworks Papers, 97.

others drew payments for maintenance and other work around the furnace. Most of their labor got expended in other branches of the ironworks; they are therefore not here counted as furnace hands.

Ten men held forth mainly in the highly skilled work of the forge and the rolling and slitting mill. John Turner and John Vinton were apparently in charge of the fineries, John Francis, of the chafery and the hammer. Quentin Pray ran the forge at Braintree. For the rest, Henry and James Leonard were skilled forge hands, Nicholas Pinnion was both a forge carpenter and a maker of iron, and Ralph Russell, Jonas Fairbanks, and Thomas Billington apparently, but not certainly, forge helpers, less experienced than the rest.[6] While these men were forge workers by main occupation, several of them carried on the related work of the rolling and slitting mill, when that unit was being operated.

Two men were smiths, occupied in the working up of wrought iron into a wide range of tools and other products. One, Samuel Harte, came as an indentured servant in Leader's time but became a free tradesman paid by the Company both for his services and the use of his tools.[7] He was probably taken on as a replacement for Joseph Jenks, another blacksmith, highly skilled, and an important figure in his own right. While, during Leader's term as agent, Jenks had undoubtedly made the transition from employee status to that of an independent, self-employed workman with a shop and at least one apprentice of his own, he must here be counted in his earlier ironworks connection. That he first worked as a hired blacksmith is clearly evident in several quarters.

The well-remunerated work of coaling occupied nine men, John Francis, when not serving as hammerman, Richard Greene, John Hardman, Thomas Look, Richard Pray, Richard Smith, Henry Stiche, William Tingle, and Henry Tucker. Richard Hood and John Parker were sawyers, Francis Perry, the works carpenter. Daniel Salmon had the operation of the farm at Hammersmith as his main concern but handled many other jobs as well. Data on four workers are so fragmentary that they cannot be fitted into any particular job

[6] Ironworks Papers, 40, 63, 69, 70, 74–75, 85, 94, 119, 122–24, 126–28, 130–35, 146–47. The payments to Pray cover forge activities in such number as to suggest that he had been the principal workman at Braintree. Important as the Leonards were later to become, they seem to have drawn only small sums for their work at Hammersmith.

[7] *Ibid.,* 54, 66, 99, 115, 158.

categories, and barely meet our criteria for full-time regular iron-works employees. These were Richard Stiche, John Dimond, John Chackswell, and John Gorum.[8] Finally there were the clerks and bosses, William Osborne, Thaddeus Riddan, John Gifford, and Oliver Purchas.[9] The first three we have already encountered. Purchas' functions, first as clerk, then as works manager, came in the post-Gifford period.

These were the free and full-time employees on whose shoulders rested the main work at Hammersmith and Braintree. Their remuneration was on a complicated basis. Managerial salaries were figured on an annual rate. For the skilled workers, however, we find payments by time period and by the piece and, within the limits of surviving accounts, with little rhyme or reason. Even in the time-period wages, there is a complication in that here there were both what the accounts refer to as "standing wages" and what appear to have been straight or actual wages. The former were low. They were apparently restricted to forge workers and, in three of four instances, to quite important ones at that. They were augmented with both term and piece wages. Paid to only four men, Pinnion, Turner, Vinton, and Russell, in amounts ranging from £1.10s. to £6, and covering from six to eighteen months, they were in each case dwarfed by these workers' total day- and piece-rate earnings. Generalizing from limited data, one is tempted to suggest that these payments were made to keep key workers on hand and their services available to the Company. If such were the case, they might fairly be called "stand-by wages."

The workers' "real" term wages, or what the accounts call simply "wages," covered employment figured, or carried in the agent's books, now by the year, now by the quarter, now by the month, now for so many weeks. In most cases, given totals are not total accumulations of day labor. By and large, they are not evenly divisible by any obvious periods of shorter duration. Of twenty-odd entries available to us only seven carry reference to particular operations. Fairbanks got paid for twenty-six weeks at the forge, Pinnion and Perry for six and nine months of carpentry, Tyler for twenty-seven weeks'

---

[8] *Ibid.*, 51–52, 55, 57, 63, 73, 77, 80, 85, 92–94, 109, 112, 115–16, 119, 125–26, 135, 143–46; Essex Deeds, I, 9, 30; LXV, 183; *Essex R. & F.*, I, 134, 136, 138, 174, 192, 198, 392, 424; II, 93, 96, 130, 193.

[9] Winthrop and Leader have been excluded on the grounds of their presumed part-owner status.

blowing and for four weeks on the furnace beam, and Wiggins and Beale for eighteen and twenty-seven weeks of filling the furnace. The sums involved were £16.6s., £4.10s., £35, £54, £3, £31.10s., and £20.15s., respectively.[10] The going rates, as best we can bring them down to a daily figure, seem to be different from, but in only one case, that of Tyler, the founder, conspicuously higher than, that which prevailed, as we shall see, for general day labor. The jobs were apparently specialized. They were probably the main occupations of the men concerned. Even among the Hammersmith "specialists," then, it would appear that it was mainly because earnings were figured at piece rates, and paid in addition to time-period wages such as these, that the yearly remuneration of some of the workers reached quite impressive proportions.

In the case of general labor, whether carried by regular ironworkers or by neighboring farmers and others, the wage structure is much simpler. For such work the going rate was 2s.6d. a day.[11] Entries in the accounts cover a wide range of activities which were so reimbursed. Unloading the boat, lathing and daubing chimneys, heaping coals, breaking up and mending the hearth, work on the hammer beam, and general labor at the slitting mill will serve as a fair sampling. Sawing probably drew the same pay. When work involved the use of teams and oxen, the day rates ran as high as ten shillings.[12] Here, obviously, more than labor was being reimbursed. In a number of individual workmen's accounts both weekly and day wage entries appear. The latter, however, always came when the men were not drawing what might be called their regular wages. By inference, when work in a man's proper field was not available, he was free to take on other jobs and get paid either by the day or by the piece.

Piece-rate jobs covered a wide range of activities, such as making a saw for eleven shillings, helving an ax and pegging a wedge for four pence, making baskets for two shillings six pence apiece, and making a roller for two shillings. More important, however, were certain regular operations which, following contemporary English practice,

---

[10] Ironworks Papers, 52, 53, 58, 62–64, 69, 70, 74–75, 78, 87, 91, 92, 94, 95, 108, 115, 118, 119, 120, 140, 142.

[11] The figure is derived by division of given money totals by numbers of days worked. Instances are so common in the accounts as to leave no question as to its soundness.

[12] Ironworks Papers, 114–15.

and without exception here at the ironworks, were reimbursed by the piece. In the woods, men earned two shillings a cord cutting wood for charcoal, a bit more when it was for use in fireplaces. For cording it, they got six shillings a score, for cleaving it, a shilling a cord. Coaling brought five or five and a half shillings per load in the years for which we have records. Getting out ore, both the bog and rock varieties, was, in general, rated at a shilling a ton. The rates on carting ranged, presumably depending on distance, between 3s.6d. and 4s.6d. per load of charcoal, and between 3s.4d. and 7s. per load of ore. Finally, the "making" and "drawing out" of iron in the forge was paid for by the piece, although the basic units are not clear in the accounts now available. In 1653, Quentin Pray got 10s.1½d. for performing the two operations on 6 cw. 2 qr. 14 lbs. of iron at Braintree. In the same year, John Turner and John Vinton drew £1.4s.10d. and £1.17s.10d. for making 1 T. 6 cw. and 1 T. 11 cw. of iron, respectively, and Nicholas Pinnion £1.14s.4d. for drawing out the iron they had made.[13]

The ironworks neighbors, the supernumeraries, were paid at the same rates, figured either by the day or by the piece, as the regular full-time workers. Most of their pay came for timber cutting and carting, although three men were paid for digging ore, eight for carpentry work, two for farm work like mowing grass and making hay, and one for mending the ways over which the charcoal was hauled. Only three were engaged in ironworking operations, two at general work in the slitting mill, one, at Braintree, for making a clevis for a coalwain. Interestingly enough, the prevailing principle of identical pay for identical work even applied to Indians. Only two, "Anthony" and "Thomas," appear in the accounts. Each got two shillings a cord for cutting wood.[14] The whole group of irregular and part-time workers, at both Hammersmith and Braintree, and over the whole span covered by our data, numbered sixty-five.

Total earnings for both types of workers, as indicated in the surviving accounts, range from the very low to the very high. In the former group were some whom we take to have been highly skilled, key workers. Henry and James Leonard, for example, are down for only £16 and £5½, respectively, and John Vinton for £22.10s.10d.

[13] *Ibid.,* 131.
[14] *Ibid.,* 84.

Billington and Diven got only £1.12s. and £2.5s. Whether this derives from gaps in the accounts or from brief ironworks employment is not clear. At the other end of the scale we have such totals as £193.9s.9d. for William Tingle, £217.2s. for Francis Perry, £182.4s.5d. for Daniel Salmon, and £173.13s.9d. for Thomas Wiggin, all for the whole span of Gifford's term as agent. The part-time workers' earnings were in most cases low, ranging from a few shillings in a limited number of instances to from £2 to £10 in most. Only sixteen drew more than £10 in the years covered by our records. Five of these worked at such jobs as tailoring, mowing and thatching, sawing and carpentry, timber cutting and fencing. The rest, and these were quite often the best paid of all the people connected with the ironworks, did the carting and hauling. Samuel Bennett, to cite but one example, received £422.15s.2d. between 1651 and 1653 for building houses, a coalwain, and a bridge, and for carting for the ironworks. We do not know his expenses. It is at least possible, however, that this close neighbor and friend of Gifford's netted more from the Undertakers' activities than anyone else, capitalists, managers, and highly skilled craftsmen in iron included.

Of one set of workmen, we cannot really tell whether they were full- or part-time employees. These were the boatmen. Six men, Theophilus Bayley, Thomas and Mark Graves, John Lambert, Ephraim Howe, and George Coales, earned from £25.5s. to £28 in 1651 and 1653.[15] Since the Company had one boat of fair size and two smaller craft, and since much of its supplies and finished products was shipped by water, one might expect a few boatmen, at least, to have been regular employees. No one of these men lived at Hammersmith. All settled at Lynn long before the plant was erected. Probably, their work situation was the marine equivalent of that of the neighboring farmers who hauled coal and ore over land, with the significant exception that it was the Company which owned the boats. If the status of the men who manned the boats is uncertain, that of another, Thomas Chadwell, who was paid for various boat-fitting jobs, is quite clear. He was an independent shipwright and not an ironworks employee.

Although from a straight totaling of earnings it would appear that certain of the part-time workers earned far more than the regular

15 *Ibid.*, 55–57, 119.

employees, we have little basis on which to indicate how either group really fared. Cost-of-living figures and comparative wage data are now available only in highly sketchy form. There is some reason to believe that the men of Hammersmith and Braintree Works drew higher than average wages for general labor.[16] In the periods for which we have records, however, it does not appear that the labor force of the ironworks was well off. So, at least, one would infer from the experience of the one worker whose individual account has come down to us with both debit and credit items, Francis Perry.

Perry's account runs from May 30th to September 29th, 1653. On the former date this worker was in debt to the Company to the tune of £100.12s.8d. By the latter his labor, and that of his sons and an indentured servant, had earned him £137.8s.6d.[17] In the interim, however, his indebtedness had increased by £47.16s.4¾d., leaving him £11.6¾d. in the red. The debits show regular receipts of wheat, malt, peas, beef, and pork and, from time to time, of such things as quantities of iron, a pair of shoes, a skillet, etc. In certain cases the Company had met his personal obligations, five shillings for the support of the minister, and various sums designated as "to Sam Archer" and "at Salem," for example. Nowhere is a wage payment in money indicated. Thus, Perry would seem to have labored, not for money but for the necessaries of life—and to have come out on the short end, even so.

Whether the supplies furnished to Perry came from a company store or from outside merchants on the Company's credit is not clear. His account, in the form in which we have it, came at a time when the Company's credit was poor. Gifford and Awbrey may have had to take on the responsibility of supplying their workmen directly, whether they wanted to or not. In an earlier period, that covered by the Riddan account fragments, the workers had been paid occasionally in money, more often by merchants' bills, not in food or supplies. If now the picture had changed, and a company store had been set up, it was probably on a quite informal basis, since neither it nor its stock got listed in the inventory of the Company's assets when financial disaster struck.

[16] *Essex R. & F.,* I, 247.

[17] Ironworks Papers, 150–55. Error by an original copyist makes his total earnings £237.8s.6d., as given on p. 150. Several credit items in this version do not appear on Gifford's general accounts which have survived.

While the period covered by Perry's account was one of stress and strain, it can hardly be argued that this man was an atypical workman who fared worse than the rest. If anything, his seems to have been a quite fortunate situation at the ironworks. Here was a man who had a cart and team, who hired an indentured servant from the Company, and whose earnings, though outstripped by his obligations, were high. A wheelwright at Salem as early as 1631, he sold his houses and land there in 1645, doubtless to move to the ironworks, with prospects of high wages during construction activities and thereafter. To the ironworks he brought not only his trade skills but, one assumes, some capital. While in many respects he has the appearance more of one of the farmer-neighbor, part-time workers than of the regular employees, his residence in Company housing puts him quite clearly in the latter category. If he wound up in the red, so, one must assume, did all the other Company workmen.

The workers presumably had enough to eat, even if they had to go into debt to secure the wherewithal. They also had their housing provided by the Company in many, if not all, instances. It is clear even in our fragmentary data that sixteen workers and their families lived in Company housing at Hammersmith and Braintree over the whole period. We know that though these houses must have been small and crude, valued at from £2 to £10, the Company paid for their erection, improvement, and maintenance. Each went under the name of the worker who occupied it. Some also provided shelter for bachelor workers and indentured servants, their regular tenants being paid by the Company for the "dietting" of such people until they were able to set up on their own. Some may also have had little garden plots. Certain workers, at any rate, grew vegetables to round out a diet which, to judge from account entries and scattered references, must have been limited and monotonous.[18]

To this point we have been dealing with the ironworkers as free wage earners. Almost without exception, however, these men had come to Lynn and Braintree in the first place as indentured servants. If at a given point they were free, it was only because they had completed their stated terms of bound service. At the same point they were working with people who were still "in service," in the restricted

[18] *Ibid.*, 51–52, 108, 126, 150, 192, 194, 197, 203–204, 213; *Essex R. & F.*, I, 136, 173; VIII, 200–203.

sense of the term. The ironworks labor force was thus a mixture of bound and free workers, with some men making the transition from one status to the other in the very years for which we have records. In this the ironmaking plants were not atypical. Their workers were no exception to Edmund Morgan's generalization that "Three centuries ago most of the inhabitants of New England were or had been servants."[19] Our data are far from complete. In all likelihood, however, no one of the regular employees, with the possible exception of William Osborne and Joseph Jenks, had emigrated voluntarily and come to the ironworks under his own power.

There was nothing invidious in the designation of "servant." True, we have come to think of the typical indentured servant as a man or woman who sold himself into service to pay off the costs of transportation to America, the service to extend over several years during which the person involved got food, shelter, and clothing but no wages. Actually, the word "indenture" means "contract," and in the terms of contract of indentured servants in general there was enormous variation depending on the individual's bargaining power and the degree of the prospective employer's need for his services.[20] Some got no pay; others were guaranteed very high wages indeed. Wholly involuntary immigrants like the Scots prisoner indentured servants exported by John Becx had no bargaining power at all. Their treatment was dictated by the consciences of the men who bought them. At the other extreme, however, an experienced workman whose skills were in great demand in both the Mother Country and Massachusetts could and did hold out for sweeping inducements to emigration. The men needed to staff the New England ironworks were doubtless in the latter category.

Of the terms of Winthrop's recruiting of the first batch of workers we are wholly uninformed. Very likely, however, he had worked out, with each man he took, a contract, oral or written, that specified what the Company would provide by way of transportation, wages, food, clothing, and housing, and what work the "servant" would perform over how many years. Subsequent additions to the labor force were

[19] Edmund S. Morgan, *The Puritan Family; Essays on Religion and Domestic Relations in Seventeenth Century New England* (Boston, 1944), 62.

[20] The interpretation followed in my text leans heavily on the findings of my former colleague, Professor L. W. Towner, in his unpublished Northwestern University doctoral dissertation on indentured labor in Massachusetts.

doubtless handled in the same way. While no copy of a contract between the Company and any of its actual workers has survived, there is reasonably good evidence showing that no less than seven employees, including several counted earlier in this chapter as free employees, made the transition from bound to independent status. The evidence ranges from a flat statement that Samuel Harte had come as a servant of the Company to payments made various individuals and indicated as coming "at going away," or at the end of their time. Others had presumably become free before the point at which surviving Company accounts begin.

That there had been a broad range of conditions of hiring is suggested by what the account books record as to the treatment of men still being carried on the books as "English servants." Three of these had been recruited in England in 1651. The Company paid their expenses until they sailed and, we assume, bore the costs of their transportation. Here they got boarded by other workmen at Gifford's charge but received no clothing, except for a pair of gloves on one occasion, and no "gifts." They did, however, earn wages, two of them getting £13.6s.8d., and the third, £6.13s.4d. a year. Another pair, Robert Crossman and Jonas Fairbanks, tentatively identified above as less experienced forge hands, were paid such high wages that they could hardly be classed as "servants," were it not for the fact that the costs of their feeding were carried in the servants' account in the agent's books. Crossman earned £36 a year. Fairbanks drew £16.6s. for twenty-six weeks' work in 1651, £20.5s. for work extending over seven months and two weeks in 1652. Similarly Jonathan Coventry, Gifford's clerk, was credited with earnings of £20.17s.9d. at a time when his diet was costing the Company £24.17s.9d., according to one claim. At the other end of the scale, Charles Hooke and William Love got boarded, clothed, shod, furnished with tobacco, and received occasional gifts. They got no wages, however, until they became free. Then they were paid like the other workmen. Other servants on whom we have less detailed information were probably handled in the same way as this last-mentioned pair.[21]

This evidence suggests that there were at least two grades of unfree labor at the ironworks, perhaps best designated as "hired servants"

[21] Ironworks Papers, 52, 55, 57–58, 60, 70, 90, 92, 94, 96–102, 107, 121, 122, 126–27, 129, 135, 146, 148, 158; *Essex R. & F.,* II, 89; VIII, 202.

and "bond servants," respectively. Crossman, Fairbanks, and Coventry were clearly of the former, Hooke and Love, of the latter sort. The hired servants had trades or skills which were in demand. To secure them the Company undertook to provide transportation, food, lodging, *and* wages. The bond servants were probably younger, unskilled, and of limited bargaining capacity. They got transportation, the necessaries of life, and nothing else, except, perhaps, the opportunity to learn a trade. The servants, in whichever group they fell, were not free. They had no choice but to serve the Company for as long as their contracts required. That otherwise they were reasonably well off, however, is suggested by evidence from several quarters.

In the first place, the clothing and other allowances to bond servants like Hooke and Love seem to have been generous, especially in the case of shoes (doubtless for good reason). The diet of all its servants cost the Company five shillings per week per man, though the Undertakers insisted that 3s.6d. should have been ample. Secondly, the services of certain of the hired servants were occasionally rented out to the regular ironworkers. The latter paid their servant helpers fifteen shillings a week, apparently at the same time that the latter were accumulating their annual wages *and* being fed at the Company's charge.[22] Finally, to shift from the account books to outside sources, there seems never to have been even a suggestion of ill-treatment of any of the Company's servants. There were runaways in England. In Massachusetts we find only one instance of legal controversy as to the duration of an ironworker's indenture.[23] All this is in marked contrast with much of the record of indentured servants in the colonial period generally.

More numerous and more interesting than these English servants at the ironworks were a group of Scotsmen who came to them for no better reason than that they had been prisoners of war. These were, without question, bond servants, utterly devoid of bargaining power, reluctant emigrants, victims of fate if ever men were. As the Parliamentary armies defeated the Royalist forces in one battle after another in the last years of the Civil War, the authorities were brought face to face with the problem of how to dispose of large numbers of prisoners, most of whom were Scots. To maintain them in England

[22] Ironworks Papers, 4, 28, 45, 138.
[23] *Records of the Court of Assistants,* III, 23.

would be costly, to send them home politically inexpedient. Though some consideration was given to their sale as mercenaries, practical politics and mercantilist considerations ultimately dictated that they be sent to the colonies.

While a number of prisoners from the Battle of Preston got transported, their destination is unknown. Of those taken at Dunbar, and there were three thousand of them, some were shipped to Ireland, some to Virginia and Barbados, and some to New England. The last were consigned to John Becx and Joshua Foote, known to us as two of the Undertakers of the ironworks. Traveling in the ship *Unity,* Augustine Walker, master, one hundred and fifty Scots crossed the Atlantic and reached Massachusetts in December, 1650. Some sixty were destined for the ironworks; the rest were to be sold off to any who who would buy them. Becx, with characteristic businessman's caution, had specified that his prisoners be "well and sound and free from wounds." A follower of Darwinian theories of natural survival might consider the specifications somewhat superfluous. Those who had managed to live through the battle and a harrowing move down from Dunbar, in which thirty men died each day, would presumably have been of better than average hardiness. No list of *Unity's* passengers has survived. However hardy, or however carefully selected, there must have been considerable mortality during a crowded, end-of-year, transatlantic crossing. Making an arbitrary but not unreasonable allowance of a 10 per cent decline in numbers, figuring £5 as cost of transportation per man, and assuming that in Massachusetts their services would have sold at from £20 to £30, Becx and Foote would appear to have turned a profit of £1500 just on this one venture alone.[24]

Another great batch of prisoners was accumulated at the Battle of Worcester which came in September, 1651, a year after Dunbar. They got similar treatment. Again Becx and Company was interested in a batch set off for shipment to New England. This time more than 270 Scotsmen embarked in *John and Sara,* captained by John Greene. Its cargo assigned to Thomas Kemble, a Charlestown merchant, the ship sailed in December and presumably reached Boston in late January or early February. While a letter from Becx to Gifford men-

---

[24] Charles Edward Banks, "Scotch Prisoners Deported to New England by Cromwell, 1651–52," *M. H. S. Proc.,* LXI (Oct. 1927–June, 1928), 4–29.

tions only 240 Scots prisoners, 272 men were listed by the Search Office at Gravesend. A copy of this list has survived.[25] Again, we can only guess at the number who survived the crossing. Again, we must assume that many did not. Again, too, we have every reason to believe that the business was highly profitable to John Becx.

Despite the connection of this key figure in the Company of Undertakers with both consignments of prisoners, there is nothing which suggests that any of the *John and Sara* Scots were destined for, or reached, the ironworks. Circumstantial evidence in a number of documents makes it all but certain that the men who wound up in Gifford's care, for longer or shorter intervals, were prisoners from Dunbar imported in *Unity*. Originally responsible for sixty-two, he still carried thirty-five on his books in the fall of 1653. Of the rest, seventeen had gone to Awbrey, three to the Company's local Commissioners, two had been sold, one had died, and four were unaccounted for. Obviously most of the human cargo of *Unity*, and presumably all of that of *John and Sara*, found its way into other hands. Quite a little is known of the New England careers of some of these victims of war.[26] Limitations of space require that we confine our attention here to the ironworks Scots.

Whatever may have been in Becx's mind, the arrival of the Scotsmen posed major problems for the Hammersmith management. They knew very little English, as the strange phonetic spellings in the passenger list and in other documents abundantly testify. They were in bad shape when they landed. Gifford had to pay out money for "physicke," for "a windeing sheet for Dauison the Scott," and for the "cure of two Scotts."[27] It was his job to quarter them. He therefore built a fair-sized house, perhaps but far from certainly, the present Scotch-Bennett-Boardman House in Saugus, to shelter some, and farmed the rest out to live with the workmen. Though the treatment accorded the Scots seems not to have been as generous as John Cotton indicated in a letter to Oliver Cromwell in July, 1651,[28] the costs

---

[25] Ironworks Papers, 28; *Suffolk Deeds,* I, 5–6.

[26] Ironworks Papers, 28–30, 33–35, 37–38, 40, 104–105, 107; Suffolk R. & F., 225; Banks, *op. cit.,* 12–14, 16, 23–28.

[27] Ironworks Papers, 103–105, 129, 149.

[28] Banks, *op. cit.,* 14. The ironworks got the largest single group of Dunbar prisoners of which we have record. Nothing in the Company accounts suggests that it provided the generous living and working conditions summarized by Cotton. If he had been

of their maintenance ran very high, as Gifford was reminded by the Undertakers in no uncertain terms. Their main grievance was the cost of food which ran Gifford five, and occasionally even six, shillings a week. This, according to John Becx, was preposterous, since in Massachusetts there was "plenty of fish both fresh and salte & pidgions & venison & corne & pease at a very cheape Rate."[29] Actually, the food charges were but part of a total expenditure on the Scots of no less than £980 between 1651 and 1653 and which included such items as the purchase of suits, shirts, shoes, stockings, gloves, and aprons in large numbers, smaller expenditures for tobacco and liquor, in part of which, at least, the Scots shared, and £35 for the framing of the Scots' house which, incidentally, was built on land which the Company did not own.[30]

What did the Undertakers get in return for this large expenditure? They got the services certainly of the thirty-five men in Gifford's care, and perhaps of those taken on by Awbrey and the Commissioners as well. Gifford's books show no credit items covering the return from the sale of certain Scots to outsiders, but presumably the Company got something in such transactions. They do indicate small cash returns on the renting out of Scotsmen to some of his regular workmen. The commonly prevailing rate seems originally to have been £5 for a man's time for two years. In taking on a Scots servant, however, a workman also took on responsibility for his food, clothing, and shelter. The Company's net gain was thus better than the rental figure indicates, at least when the workmen met their obligations.[31] All in all, though, the bulk of the return on the investment in the Scots came in the form of labor for the Company, labor for which the Undertakers would have had to pay at high rates had not the prisoner-servants been on hand.

Many people have wondered why the Scotch House was built so far from the ironworks proper. The answer is comparatively simple. These indentured servants worked far more on the farm and in the woods than at furnace and forge. There is no questioning the fact that the Scots' labor was primarily unskilled. We saw earlier that in

correctly informed, there must have been someone who took on the services of an even larger number. I have found no clue as to his identity.

[29] *Essex R. & F.,* II, 89; Ironworks Papers, 38.
[30] Ironworks Papers, 43, 58, 126–30, 148–49; *Essex R. & F.,* II, 89.
[31] Ironworks Papers, 45, 138, 153; *Essex R. & F.,* II, 96.

1652 and 1653 these men took over what seems to have been the greater part of the woodcutting job. They also did a certain amount of mining. In all probability, too, they did most of the farm work, making hay, growing corn, etc. In some of even this more or less crude work the Scots had had to be taught how to proceed.[32] For more specialized jobs far more training was required and, to judge by presently available evidence, precious few managed to acquire it.

In Gifford's accounts only two Scots appear to have become really skilled workmen. One, John Clarke, was a smith to whom Samuel Harte taught his trade, at a cost to the Company of £4 in 1653. Another, Thomas Kelton, seems to have been both a collier and a miner. Where and how he acquired the training for the former work is wholly obscure. Clearly, the craft apprenticeship of others was in process at Hammersmith. Gifford, we know, in protesting the valuation of the Scots at only £10 per head in the 1653 inventory, claimed that besides Clarke, who would save the Company £44 a year in smith's wages, there were three forge workers, a hammerman whose work would eliminate an expenditure of five shillings per ton of bar iron, two carpenters capable of handling all but "some work extraordinary," and six colliers whose labors would permit the Company to dispense with the services of all but a single "M(aste)r Collier."[33] Almost certainly, however, the agent was exaggerating. Training in the indicated fields may have been fairly well along in these relatively few cases. It can hardly have been as nearly complete as Gifford was suggesting.

Regular ironworkers, neighboring farmers and tradesmen, English and Scots indentured servants—these were the men who staffed America's first successful ironworks. Their jobs, wages, and living conditions outlined, we may turn to consider them as people. We have already mentioned their deviations from prevailing Puritan standards, which posed problems for managers and magistrates. To the workers themselves such brushes with the law probably were counted as some of the costs of living with the all too godly Puritans, whose religious convictions they did not share. The workers were in their colony. Conformity was a *sine qua non,* like it or not, and, given the discrepancies between the patterns of conduct of a rough lot of iron-

[32] Ironworks Papers, 9, 46, 77–79, 95, 158.
[33] *Ibid.,* 9, 115, 117, 126, 252–53; *Essex R. & F.,* II, 96–97; Suffolk R. & F., 225.

workers and of those bent on raising God's kingdom in Massachusetts, rubs and even explosions were to be expected. Here, in essence, was an early version of the phenomenon of assimilation of immigrants into a settled "American" community, a process which started as soon as the thin lines between "old" residents and newcomers could be distinguished, and has carried right down to today.

One word of caution is called for. Modern Americans have become the victims of a stereotype of the Puritans which makes them so good and proper as to be less than human. Obviously the Puritan social ideal was a high one. Equally obviously, there was a discrepancy between the ideal and the real, as a quick scrutiny of the early Massachusetts court records will readily demonstrate. Every offense of an ironworker can be matched by one of a regular inhabitant. True, the authorities looked on the ironworkers' lapses as special problems. Winthrop had had to intervene in their or the Company's behalf, and the Undertakers had seen fit to apologize for the difficulties their men had caused. Since the sins of the men of Hammersmith and Braintree seem not to have differed, at least in kind, from those of others, one must wonder if the magistrates' special sensitivities did not reflect a clear in-group—out-group situation. To them and to all the Puritans the ironworkers were "different." And the lapses of those who are different always loom larger than the failings of "our own kind."

Among the offenses of the ironworkers, and of their wives, swearing and cursing and assault were the most common. The former appear in the court records as presentments for swearing or for "common swearing." Occasionally, the charge is more specific, as in the case of the Quentin Prays, man and wife, who were fined fifty shillings "for five oaths," or in that of Nicholas Pinnion who was alleged to have sworn "by God's wounds" and "by God," or in that of John Hardman who was ordered to be fined and whipped for "Many horrible oathes and many filthie unclean and wicked speeches."

Several of these verbal lapses accompanied acts of violence which came under the heading of assault. John Hardman, for instance, also got fined for "breaking the head" of a Salem resident whom he encountered on the highway. Quentin Pray achieved similar results with a rather vicious instrument, "a staff having an iron two feet long on the end of it," his victim being a fellow worker, Nicholas Pinnion. Two men, one a servant, committed acts of assault and had their estates restrained in consequence. Wife beating brought a number of workers before the courts, not infrequently with an attendant airing of lurid details.

The Prays and the Pinnions in particular seem to have had serious marital troubles. In January, 1648, it was deposed that Richard Pray had been heard to call his wife a jade and a roundhead, and announce that he would beat her twenty times a day before she would be his master. When she objected to his swearing, he allegedly "took up a long stick about the size of the great end of a bedstaff" and swung at her. The deponent warded off the blow but Pray managed to kick her against the wall despite his efforts. On another occasion, and with the same provocation, he "took his porridge dish and threw it at her, hitting her upon the hand and wrist, so that she feared her arm was broken." Finally, in the witness' own words:

Some one present told Pray that the court would not allow him to use his wife so, and he answered that he did not care for the court and if the court hanged him he would do it. It was said to him that the court would make him care, for they had tamed as stout hearts as his, and Pray answered that if ever he had trouble about abusing his wife, he would cripple her and make her sit on a stool, and there he would keep her.

All of which sounds serious but it is as nothing compared to a couple of cases where we are dealing with major incidents of violence. One stemmed from the Pinnions' stormy married life. The lady, apparently smitten with another man, made the somewhat astonishing charge that her husband, Nicholas, had killed five children, one of them a year old, and had also so beaten her that she had miscarried, "being a quarter gone with child." In the other the formal designation of the crime reads as follows:

John Turner, living at the iron works in Lin, presented for stabbing Sara Turner, his daughter-in-law, and swearing by the eternal God that he would kill John Gorum, and for being overtaken in drink, etc., to be severely whipped at Salem; then to be sent to Boston prison until he be whole; and later to be whipped at the iron works.

This sentence was revoked. Unfortunately we are not told why.

If swearing accompanied some of the assault cases, so, as in that just mentioned, did drunkenness. With little question, this failing was in the background of many of the crimes and misdemeanors which are so vividly portrayed in surviving court records. Only ten people seem to have been formally charged, at one time or another, with being drunk or overcome with drink. Liquor also figured, however, in cases involving cursing and swearing, slander of fellow workers, contempt of magisterial authority, etc. Absence of its mention in others which reached levels of ribald frankness too flavorful to present here is much to be wondered at. If the ironworkers, without alcoholic stimulation, reached the heights of unbridled statement and animal spirits which the Essex County Court Records and Files suggest in their published, and document in their unpublished, form, truly they must have been a rowdy lot, a very plague upon the magistrates.

Life at Hammersmith and Braintree Works obviously had its Rabelaisian side. The printed *Records and Files* indicate only one sexual lapse by an ironworker. Here a warrant was issued in the case of a young man charged with "suspicion of uncleanness with Jane Somers." Buried away in manuscript, however, are data indicating fornication and adultery, the coveting of neighbors' wives, and incidents of gross sexual play. Pinnion, for example, once ordered a man out of his house. His wife announced that if he went she would go also. On another occasion he had suggested that she had taken one of the workers to bed. Sara Turner boasted that she could freely dispense her favors and never be "mistrusted for a dishonest woman." And Dorothy Pray but narrowly escaped rape by one of the ironworkers in John Hardman's house and during the latter's absence. These are but a few of numerous instances.

Loose tongues were another source of trouble. The ironworkers' wives were not infrequently haled into court for such offenses as "scolding" and "speaking opprobrious words to their neighbors." The spouse of Richard Pray, to cite but a single example, was convicted in 1650 "for that she should say to her mother in lawe get you whom yow old hogge get you whom and withall threw stones at her...." The women had no monopoly on verbal lapses, however. Men also thus sinned, and among their offenses those involving contempt of the magistrates are perhaps the most significant, if not the most pungent. We have already seen what Pray is alleged to have thought about the Massachusetts courts. Another worker was charged with "slighting of authority." He had dared to question the justice of one of the judicial findings in the legal proceedings in which Gifford became embroiled after the Company's collapse. When two young men were on trial, one for the attempted rape mentioned above, another for wanton behavior, testimony indicated that one of the culprits had said that the magistrates were "more devils than men" or something to that effect. Although it was pointed out that if this were brought to the attention of the authorities, the offender's punishment would have been stiffer, he seems to have gotten away with his statement. Clear proof was doubtless lacking. In general, however, the courts seem to have been surprisingly forbearing in such matters.

All of the crimes and misdemeanors so far cited could have taken place in any society. The ironworkers were also guilty of some which were largely peculiar to mid-seventeenth-century Massachusetts, absence from church, violations of sumptuary laws, and criticism of churchmen, for example. The workers ducked church frequently, and often made things worse by drinking and playing when they should have been at divine service. Among the offenders in this area were Thomas Beale, Nicholas Pinnion, Henry Stiche, who drew only an admonishment for "coming to meeting not once or twice in a year," and Joseph Jenks. In November, 1652, Nicholas Pinnion and his wife, the wife of Joseph Jenks, jr., John Gorum, John Parker, and Richard Green were all fined for wearing silver lace. A daughter-in-law of Francis Perry was similarly punished for wearing a silk hood. Jonas Fairbanks but narrowly escaped a fine for "wearing great boots" by demonstrating that "he did not wear them after the law was published." Though here, too, there were offenders who were not ironworks folk, it is safe to infer that the latter, male and female, were bent on dressing as well as their means permitted, and better than the law allowed. Finally, in the matter of criticism of the clergy we find only a single incident. That one, however, is quite interesting. Thomas Wheeler arose in Lynn town meeting in February, 1654, and made "Euell and Sinfull & offensive Speeches against Reuert Teacher Mr. Cobbett in Comparing of him unto Corah. . . ."[34] The

[34] These various cases are documented in the following sources: *Essex R. & F.,* I, 107, 130, 133–36, 138, 151, 156, 173–74, 183–84, 205, 271–72, 274, 360, 379, 393, 414; Essex R. & F., I, 90, 112, 114; Ironworks Papers, 54, 75, 80, 86, 96, 105.

circumstances are not clear. If, however, an ironworker had in fact drawn an analogy between a respected clergyman and the Biblical figure symbolizing challenge to the privileges of a divinely ordained priesthood, he was speaking out of turn as well as pinning on Mr. Cobbett a name which, in Puritan eyes, ought better to have been applied to the audacious ironworker himself![35]

If we look at all of the ironworkers' offenses over the whole period, a certain patterning seems to emerge. When the plants were being erected, there was no trouble from the workers, whether because they were numbed by the strangeness of their new environment, or too busy to get into difficulty, or because the Massachusetts officials, anxious to get the works erected, saw fit to look the other way from time to time. The recorded lapses, at any rate, do not appear before November, 1648. Then they turn up with considerable regularity. In Leader's later years as agent the situation was clearly bad. In Gifford's time, there was improvement but the workers were by no means absent from the court sessions, even so. Apparently, once Hammersmith was in production, the workers went off the deep end *or* the magistrates had decided that the time had come to deal with them as with other lawbreakers. Then, either because the treatment took hold, or because Gifford kept the workers more or less in line, or because production slumps got reflected in lower earnings and, one assumes, in less alcohol, comparative peace reigned. It was only after November, 1655, however, that the workers settled down to orderly patterns of conduct. From that time only four lapses are on record, one in 1657, 1659, 1660, and 1663, and none of these was serious.[36] One's first impulse might be to say that the ironworkers had been finally assimilated or whipped into line. This was doubtless true in part. The real tapering off came, however, at a period when, as we shall shortly see, production was fitful, or worse. Some workers had gone on to other places. For those who remained, there could hardly have been the impetus toward gay and riotous living, or the wherewithal that made it possible, which had prevailed when Hammersmith was a really going concern.

Except for the indirect evidence of less frequent appearances in court, one might fairly wonder if the ironworkers had been anything

[35] Lewis and Newhall, *op. cit.*, 236–38.
[36] *Essex R. & F.*, II, 35–36, 167, 196; III, 83.

but an alien element in Massachusetts during the years in which the Hammersmith and Braintree plants were in operation. Take, for example, the matter of admission to the ranks of freemen. Many of the part-time workers were admitted to voting privileges. Some, indeed, had made the grade prior to their ironworks association. Only three men here classified as real ironworkers seem to have become freemen, and no one of them during the years in which the works were running.[37] Since the privilege depended on church membership, and since this depended on evidence of conversion, the situation is hardly strange. Similarly, it was only after Hammersmith had been abandoned that ex-ironworkers began to serve on juries or as constable. Only two achieved even this.[38] Thus, so far as we can tell, the really effective assimilation of ironworkers came late, and often at other places. Even there, as we shall see, the process was considerably less than easy.

Within Hammersmith itself only one worker, Joseph Jenks, seems to have made a real mark of his own. Though atypical in many ways, has story cannot be omitted here. This skilled worker was, as we have seen, a blacksmith. Probably recruited by Winthrop in Maine, in the course of his iron prospecting expedition, he may have worked in the setting up of both Braintree Furnace and the Hammersmith plant.[39] Originally an employee of the company, paid £13.8s.6d. between October, 1648, and October, 1649, according to the Riddan account fragments, he became the independent or semi-independent operator of a toolmaking forge. Prior to the dates of the account entries just mentioned, as early as January, 1648, in fact, he obtained from Leader "libertie to build & erect a mill or hamer for the forging and making of sithes or any other ware by water at the taile of the furnace & to have full benefit of the furnace water when the furnace goes provided he damnifie not any works that may hereafter be erected. . . ." Construction was to be finished by June 24th, 1648, on penalty of re-entry by Leader. The agent, besides making the site available, furnished the iron needed in setting up the mill.

[37] Savage, *op. cit.,* II, 457; III, 319; *Essex R. & F.,* IV, 38.

[38] *Essex R. & F.,* III, 263; IV, 431; VI, 215; VII, 398; Savage, *op. cit.,* III, 394; Lewis and Newhall, *op. cit.,* 577–78, 580.

[39] I have found no evidence linking Jenks with ironworks activity at Braintree. The earliest references to him in Essex County is as party to a lawsuit dated December

So far as we can tell, the Company was thus farming out to an enterprising employee an iron-fabricating operation which it was unable or unwilling to undertake on its own. Whether the little mill was wholly Jenks' or partly his, partly the Company's, will probably never be known.[40] Clearly, however, Jenks worked it, buying considerable quantities of bar iron from the Company and converting it into tools and other artifacts. The range and calibre of his operations are attested by both documentary evidence, such as his receipt of industrial patents from the General Court, and archeological findings at the hammer-mill site. The latter include not only nice examples of wrought-iron craftsmanship but specimens of brass plating, and hundreds of brass pins. They seem to indicate a small plant consisting of at least a forge hearth and a power-driven hammer of fair size.[41]

Unlike the other Hammersmith workers, he owned his own house and a gristmill, items which the Company apparently purchased from him in 1652. How these had been acquired in the first place is nowhere indicated. While the Riddan accounts show payments to Jenks for work done by men clearly identifiable as regular Company employees, this able craftsman had at least one apprentice of his own. William Curtis, to whom Jenks so well taught his trade that he could be recommended to Winthrop at New Haven as a qualified smith, was, with little question, Jenks' and not the Company's servant. It is possible that he had other employees as well, a son of the same name, who was later to be connected with ironworks at Concord and Pawtucket, doubtless having been one of them.

How well Jenks fared in all this is not at all clear. The reasons for the sale of his house and gristmill are wholly obscure. So are those for his mortgage of his "forge, working houses and works with all the appurtenances thereunto belonging," to Gifford in 1651. One suspects that he was short of money. Certainly, he did not derive much from the Company. Gifford's accounts show payments to him of only £5, for tools and other equipment, except for the purchase price of house and gristmill. Other obligations to Gifford are also

28, 1647. (*Essex R. & F.,* I, 130.) By then, it is safe to assume, he was living and working at Hammersmith.

[40] Essex Deeds, I, 22, 32.

[41] *Mass. Records,* II, 149; III, 386; IV-1, 233. The Jenks plant has not been restored. Portions of water wheels, anvil blocks, framing, and countless metal artifacts found in the area are on display in the Restoration Museum.

indicated. In the strained period following the Company's collapse, Jenks picked up some of its assets, the slitting mill as well as the property that had originally been his own. These items, too, he was mortgaging to Simon Bradstreet in 1657. Though Jenks seems to have lived at Lynn until his death in 1683, it is doubtful if he had managed to make a financial success of a career that included admirable technological accomplishments and that transition from employee to entrepreneur which has so often been repeated in American industrial history. He nevertheless deserves a hearty "Well done" from posterity.[42]

Jenks was not the only ironmaker left at Lynn when Hammersmith, like its Braintree branch plant, got abandoned. He presumably continued to ply his smith's trade. Others exchanged their ironmaking skills for farming. This happened with Samuel Harte, John Diven, Daniel Salmon, and Richard Hood, and doubtless in the case of others whose later careers are lost in obscurity.[43] Some, of course, moved on. Certain workers went to other ironmaking establishments. As the real links between the plants at Lynn and Braintree and the later American iron industry, their story is here postponed to the final chapter. For certain others it is not clear whether a change of residence meant a change of occupation. Quentin Pray, for example, stayed at Braintree until his death in 1677, and John Francis moved there. Whether either or both worked at a later Monatiquot River forge is not indicated. Richard Pray went to Providence along with Mary, his ill-tempered wife, and may or may not there have earned a living as a collier. In other cases a change of vocation may be taken for granted. John Chackswell left for Barbados while Hammersmith was still a going plant.[44] Francis Perry was in the same place before the summer of 1655.[45] John Diamond moved to Kittery, Maine, prior to 1652, probably to work for Leader, by then occupied with sawmills. Between 1657 and 1660, Thomas Wiggin settled in Rustport, New York, although we are not informed as to either his motives or his

[42] Jenks' activities and transactions are documented in the following sources: Ironworks Papers, 90–91, 96, 140, 142; Essex Deeds, I, 22–23, 33, 36; V, 65; Savage, *op. cit.*, II, 543; Lewis and Newhall, *op cit.*, 208; *Essex R. & F.*, II, 210.

[43] Essex R. &. F., XL, 87; Essex Deeds, II, 127; III, 2, 37; VI, 37, 54, 111; VII, 31; IX, 12, 238; XXV, 218; XXVI, 203; *Essex R. & F.*, II, 36, 338; III, 268; V, 256.

[44] *Essex R. & F.*, II, 193.

[45] Essex Deeds, I, 62.

eventual occupation.[46] Others, disappearing from local and county records, must be assumed to have gone on to other places and become lost to history. Data are so fragmentary, however, that an attempt to determine the proportion of exodus from ironmaking to other callings is wholly impossible.

If with the exception of certain families, the Leonards and the Jenkses especially, the men of Hammersmith and Braintree Works and their immediate descendants left little mark in the communities in which they lived, neither did they leave much of the world's goods behind them when they died. William Osborne, no ordinary workman, managed to accumulate an estate of £836.7s.5d. by the time he died in August, 1662. A man at the other end of the ironworkers scale, James Moore, a Scots servant, left assets of some £56 when he died about 1660. John Francis, deceased in October, 1668, left but £18.10s. in goods and £25.4s. in debts due his estate. And Samuel Harte died possessed of £156. Theirs are the only wills and inventories that seem to have survived.[47] The small number suggests that most of the ironworks people passed on to their reward in less than even modest circumstances.

The workers' legacy of children is far more impressive. Our data on the workers' families are not extensive. Of eight of the full-time employees, we are not even certain that they were married. The others, however, produced families which were large by modern, but by no means unusual according to colonial, standards. Robert Crossman, for example, was the father of ten, Richard Hood and James Leonard, of nine, John Vinton, of seven, and Nicholas Pinnion and John Turner, of five children of whom we have record. John Diven, however, had but one son; John Hardman, a son and a daughter; and John Turner, two sons. Some had been born in England, some as fathers worked at the plants, and mothers carved out homes in their crude Company houses. More, however, appear to have been born after their fathers severed their connections with the ironworks. These findings must be regarded as tentative, given the carelessness in the recording of births which was all too common, even among settled residents. Were good records available, a wholly different picture of the size of the worker family might emerge.

[46] Savage, *op. cit.*, II, 50; *Essex R. & F.*, II, 35; Essex Deeds, I, 80.
[47] Suffolk Probate Records, IV, 106–107; V, 132; *Essex R. & F.*, II, 215–16; Essex R. & F., XL, 87–1.

Of the children, as children, we know absolutely nothing. We assume they lived much as did those of the regular inhabitants. Once grown, it was a matter of early marriage for the girls and of learning a trade for the boys. While our data on the marriages of the ironworks young people are, again, disappointingly thin, there seems to have been no special tendency toward marriage within the group. There was not a single marriage between the children of men whom we have been regarding as full-time employees. In only eight instances did sons and daughters of ironworkers marry into families having even a remote connection with the plants or with ironworking trades in general. The closest thing to a real Hammersmith union was the marriage of a daughter of Nicholas Pinnion to James Moore, an ex-Scots indentured servant. There were five marriages between sons and daughters of ironworkers and those of Lynn and Braintree farmers who were part-time employees. Finally, in two instances workers' children were wed to members of families associated with later ironworks but which, so far as we know, had no connection with Hammersmith and Braintree. All of the other marriages of which we have record were with "outsiders," with members of the community at large—which may serve as a better index to the degree of ironworker assimilation than any of the evidence cited earlier.[48]

The learning of trades by ironworkers' sons seems to have tended in the opposite direction. For those on whom we have data, at any rate, it looks as though the skills of fathers had been passed on to their sons. Two sons of Joseph Jenks, three of Henry Leonard, five of James Leonard, three of John Vinton, two of John Turner, two of Nicholas Pinnion, and one of Francis Perry, Quentin Pray, and Ralph Russell all worked in one phase or another of the manufacture and processing of iron. There is nothing with which to document the formal apprenticeship of any worker's son. In the surviving Gifford accounts, the work of only two sons seems to have been reimbursed in wage payments to their fathers, one of Francis Perry's for carting, one of Quentin Pray's for mending dams and floodgates.[49] How the ironworks trades proper were acquired by the various young men is therefore wholly open to conjecture. From actual evidence,

[48] Savage, *op. cit.,* I, 39; II, 29, 53, 136, 368, 457; III, 79, 438, 503; IV, 277, 375, 619.

[49] Ironworks Papers, 95, 113, 115, 135, 151.

however, the reservoir of iron craftsmanship represented in the training of even the men just listed must stand as one of the richest legacies of the ironworkers, as of the ironworks themselves.

All of which is hardly a recording of rich, full lives of fascinating human beings. Surely, however, our inability to make these people seem human in anything but their misdemeanors and shortcomings reflects the inadequacy of our sources, not the deficiencies of the people themselves. These were the little people, the ones to whom history is never kind. They may have been singers of songs. We know nothing of them. They must have faced the inevitable human crises of birth and love and death. Their reactions to them are lost forever. They worked, we assume, to make things better for their children, but of their successes and their failures here we are almost wholly ignorant. In their own story, on average, failure and disappointment, not all of it of their own making by any means, seem to loom larger than accomplishment and satisfaction. In that of some of their progeny the picture is quite otherwise. If Winthrop, the first agent of the Company of Undertakers, became governor of Connecticut, a grandson of one of his workmen, Joseph Jenks, became governor of Rhode Island.[50] If at Hammersmith and other seventeenth-century ironworks the Leonards were still in shirt sleeves, certain of their descendants were, on the eve of the Revolution, among the first families of Taunton, Massachusetts. More or less similar items could be cited *ad infinitum* were we here to trace out the genealogies of American families which go back to the humble folk of Hammersmith and Braintree Works.

Though in almost every instance the road to better things was to be traveled by their progeny, it was these little people who, for one reason or another, had seen fit to take the risks of emigration, to endure the strains of life in a still primitive community, to adjust to a New World and a new way of life. Surely some of the attitudes, the values, the hopes and ambitions of the men who made the bridge from English to New England ironworks were passed on to their descendants. If the trade was, on the whole, amazingly conservative, this did not stand in the way of Jenks' inventive bent, nor in that of the unknown people who first saw that the Nahant rock ore could serve as furnace flux. If the ironworks became business failures early in

[50] Savage, *op. cit.*, II, 543.

the game, its employees, during operations by the Company of Under-takers and by Massachusetts merchants to whom the courts handed over the plants, had earned a living, developed their trade skills, raised up sons and daughters, and adapted, more or less successfully, to a strange environment. Even as mirrored in the sparse data now available, these men, the prominent and the obscure, the skilled and the unskilled, well deserve our recognition as pioneers in the working ranks of American industrial capitalism.

11

A Plague of Lawsuits
and a Bankruptcy

As the country had hitherto begun to flourish in most English manufactures, so liberty was this year granted to make iron; for which purpose a work was set up at Lynn, upon a very commodious stream, which was very much promoted and strenuously carried on, for some considerable time; but at length, whether *faber aut forceps, aut ars, ignara fefellit,* instead of drawing out bars of iron for the country's use, there was hammered out nothing but contention and lawsuits, which was but a bad return for the undertakers; however it gave the occasion to others to acquaint themselves with that skill, to the great advantage of the Colonies, who have, since that time, found out many convenient places where good iron, not much inferior to that of Bilboa, may be produced. . . .

Hubbard's *General History of New England.*[1]

It is time to pick up the later strands of the ironworks' story, to turn to the "contention and lawsuits," for which this vigorous contemporary statement stands as apt text—if not as epitaph for the whole pioneer industrial enterprise with which we are concerned. Here the business and administrative difficulties, which we have seen in germ, flowered in what was for all practical purposes a bankruptcy. Then the financial collapse generated a second growth of snarls and troubles, all aired in endless lawsuits in Massachusetts and English courts. When the tumult and shouting died, a full generation later,

[1] *2 M. H. S. Coll.,* VI, 374.

the ironworks which had been set afoot with such high expectations were extinct, their achievements and failures marked only by what lingered in men's memories, the ruins of the Lynn and Braintree plants, and literally bulging court records and files.

Failing to receive adequate "accountings," the Undertakers probably lacked a full picture of the modest production, heavy costs, and disappointing sales. Clearly, however, they were not wholly in the dark. There must have been at least some letters from Awbrey and Gifford. However eloquent, they could hardly have conveyed, short of perjury, an impression of healthy business operations. The bills of exchange which their agents were drawing on the English capitalists were landing at their doors, as the bar iron with which to cover them was not. Shareholders resident in Massachusetts could document the Company's grim situation, and would certainly have let their London associates know how things stood. Joshua Foote, for example, well versed in the iron trade, was in Boston by 1652. His letters to the Undertakers have not survived. There must have been some. They must have been received as the opinions of an investor who knew whereof he spoke. Finally, we may infer that there had also been word-of-mouth reports from New England immigrants who, for one reason or another, returned to England. John Turner, the finer, to mention only one, had apparently told John Becx considerably more than that he wanted to switch to the better paying collier's job when he went back to Massachusetts.[2]

In the spring of 1652 the Undertakers announced in a letter to Gifford that four New England residents, Robert Bridges, Joshua Foote, Henry Webb, and William Tyng, were to be set up as their "Commissioners," with absolute power to act in the Company's behalf. It was to be their job to straighten out the ironworks business problems. To accomplish this, they were given a power of attorney, drawn at London on April 16, 1652, which authorized any three of them to serve as "True & lawfull deputies & attorneys," to obtain a financial accounting, not only from Awbrey and Gifford, but from Leader and Osborne as well, to secure an inventory of all stock and wares, and to assume supervison of the whole business undertaking. Under this instrument the commissioners came to act as *de facto* owner-managers. Whether by intent or otherwise, the Undertakers

[2] Ironworks Papers, 30, 35.

had evidently inserted a new level of control between themselves and their old agent and factor. To Webb, Foote, and Tyng now went part, and probably the larger part, of the ironworks output.[3] From them came, or should have come, the supplies of money and materials with which to keep the plants going. Gifford and Awbrey, to this point all but free agents, thanks to the communications block which was the Atlantic Ocean, were now under the thumbs of on-the-spot bosses. They were little different from hired clerk and foreman.

The transition was as strained as it was sweeping. While the Undertakers' letter suggests that it was Awbrey who was really on the spot, the commissioners being authorized to replace him, Gifford hardly welcomed his new supervisors with open arms. When Bridges, Webb, and Foote turned up at Hammersmith to get an accounting from him in April, 1653, the agent flatly refused. He was not going to be their "Jacke Boye," he announced. Eventually, he agreed to let Webb's clerk copy items from his books, and to provide an account of the stock and iron on hand. It took much "presing and perswayding" from the Reverend Mr. Cobbett, teacher of the church at Lynn, however, before Gifford would even allow the commissioners to inspect the iron stored in his warehouse! How the clergyman came into the picture is not clear. Since the commissioners had been "in business" at least as early as September, and since all of the interval can hardly have been taken up with a survey of Awbrey's records, one might surmise that the above had not been the first encounter between the Undertakers' commissioners and agent. Perhaps the former had had reason to conclude that, however legitimate their request, they would do well to have a blessed peacemaker on hand as they made it.[4]

The state in which the commissioners found the agent's and factor's books is impossible to estimate from this distance. Their discrepancies and discordances provided much of the fuel of much subsequent litigation, but such accounts as have survived are fragmentary and, by definition, controversial. That they had found the ironworks affairs in general in as bad shape as the Undertakers feared is incontestable. By the time they asked Gifford for an accounting, they were much worried. Indeed, they insisted that they wanted to

[3] *Ibid.*, 38–41; *Suffolk Deeds*, I, 229–30.
[4] Ironworks Papers, 176–77; *Essex R. & F.*, II, 75.

inspect the stock and products on hand in order to find out "whether thaye had wher with to Cary on end the worke."[5]

No bookkeeping could conceal matters of public knowledge and public record, the debts incurred to keep the works in operation. In June, 1652, Awbrey had mortgaged the ironworks, and bound himself in the sum of £1500, to guarantee payment in London on bills of exchange drawn to cover the claims of Jeremy Houchin, a Boston tanner, claims which totalled less than £830. In the same month he mortgaged his own "Interest, Right & title" in the ironworks, and incidentally in no other quarter do we find the slightest suggestion that he had one, to cover similar bills of exchange drawn to the credit of Joseph Rock, a supplier of clothes and other goods to the ironworks. If the Company refused payment on these bills, Rock was assured 30 per cent damages. This transaction involved only £200 but, even as it was entered into, Awbrey was then already £125 in debt and was continuing to draw supplies. Clearly, the credit picture was not good.[6]

All this must have disturbed the commissioners. They soon found it necessary, however, to take similar steps of their own. Soon after their visit to a recalcitrant John Gifford, three of the commissioners, Bridges, Foote, and Webb, mortgaged the Company's entire assets in order to secure the claims of Webb, Houchin, and Gifford. According to the instrument, Webb was owed £800 or £1000, and Houchin, £555. The amount due Gifford was not specified. While the commissioners or their successors were assured two years in which to pay off the debts, failure so to do would have seen the ironworks handed over to a commissioner, a supplier, and an employee, respectively.[7] If Webb was seeing to it that *he* got paid, whatever happened, Gifford was apparently bent, this early, on recovering from the commissioners the sums he had presumably expended for the Company, and for which he was holding his new supervisors responsible. Even in this one document, the twin roots of a gigantic legal maze—the claims of the Company's creditors and the feud between its commissioners and its agent—are implicit and intertwined.

---

[5] Ironworks Papers, 176, has "cary *or* end the worke" quite clearly a misreading of the "on" of the MS.

[6] *Suffolk Deeds,* I, 228–32.

[7] *Ibid.,* I, 306–307.

Though Webb and Gifford were "partners" in this mortgage, they were otherwise utterly at odds. Bankruptcies, present or prospective, as well as politics, make strange bedfellows. Webb saw the ironworks as in "a sore languishing state," and chalked most of this up to Gifford's "improvident husbanding." His accounts were a mess. Debts greatly outweighed assets. Even the workmen were unpaid and had done nothing for weeks at a time "for want of breade." Reluctant to reach a proper accounting with the commissioners, Gifford was no laggard when it came to calling on them for supplies. At the same time he declined to honor their orders to send the iron with which to meet the creditors' claims. Money was hard to borrow, and the more so as bills of exchange came bouncing back from London. Creditors, long unpaid, were asking for security, and not only was there no stock or "effects," but some, including the magistrates, doubted if the commissioners actually had the right to mortgage the physical plant.[8]

Gifford offered a wholly different picture. In one of his almost innumerable petitions he assured the General Court of Massachusetts that he had worked well and faithfully and had brought the ironworks to such flourishing state that his principals might have "tasted of the sweetness" of profit on their investment if it had not been for the commissioners. He had delivered his employers' estate to them as instructed. They, however, though often beseeched, had failed to work out a settlement and accounting with him. The agent insisted that, "upon a due search and examination," it would be plain that the Company owed little or nothing, that what had been delivered in supplies had been properly covered by shipments of iron or bills of exchange. Finally, he implied, and by no means obliquely, that it was Awbrey and the commissioners who had brought things to such a sorry pass.[9]

The basic questions are quite simple. Was the Company on the verge of bankruptcy? If so, to whom was the business collapse attributable? The former was quickly settled by the courts but, emphatically, not easily and to everybody's satisfaction. The latter question got different answers at different times and in different quarters. In their determination the full judicial apparatus of seventeenth-century Massachusetts was called into action. There were suits

[8] *Essex R. & F.*, II, 75–80.
[9] Mass. Arch., 59/52–55.

and countersuits, decisions and appeals, the proceedings full of irrelevancies and reeking with a spirit of contentiousness that would never call it quits. If the ironworks difficulties produced questions difficult or impossible to solve, it was thanks to their complexities, and the vigor of the contending parties, and not to defects or deficiencies in the judicial system.

That justice was not administered crudely or on an *ad hoc* basis in the Bay Colony in the second half of the seventeenth century is beyond question. Court structure and court procedures were well established by the time they were strained to the limit by the welter of ironworks litigation. Limitations of space rule out extended description here. Without a brief sketch of the prevailing system, however, the maze of proceedings we are about to examine would be close to meaningless to all but legal historians. Civil and criminal cases of at least moderate importance were normally begun in the county courts. These were presided over by a multiple judiciary properly consisting of no less than five magistrates (governor, deputy governor, and assistants) and other men of substance, although three qualified persons, always including at least one magistrate, could constitute a quorum. Trial was by jury composed of men drawn from the towns. These courts met quarterly, that of Essex County, where most of the ironworks cases were heard, meeting at Salem in its first and third quarter and at Ipswich in its second and fourth sessions. At the same level as the county courts came the "Special," or "Strangers'," or "Merchants' Courts." An act of 1639 gave "strangers" the privilege of calling a "special court" at any time to provide quick settlement of pressing mercantile problems. Three magistrates were sufficient here, their competence identical with that of the judges of the county court, and, again, trial by jury was mandatory. Since the whole purpose of the Strangers' Court was to render justice promptly, there was no appeal from its findings.

Appeals from county courts, and original cases involving grave matters excepted from their jurisdiction, were heard by the Court of Assistants, which met twice a year at Boston. Finally, at the peak of the structure came the General Court itself, the highest authority in the Colony, and a body which combined legislative and judicial functions. The latter were on an appellate basis. The General Court heard appeals from the Court of Assistants in certain cases. More commonly it pondered cases handed up from lower courts, cases in which, whether in consequence of disagreement between magistrates and jury or otherwise, no settled conclusions had been reached. By the mid-century point, neither it nor the Court of Assistants heard any cause which had not had both a trial and a review before a lower court. Similarly, both the General Court and the Court of Assistants handled their cases wholly on the basis of evidence presented in the inferior judicial agencies.

An oversimplified summary of a typical civil action might help the reader to follow the intricate course of ironworks litigation. An injured party, let us say, lodged a suit at a county court. The action, of debt or of case,[10] was entered on the first day of court. Summonses and attachments were served by the constable. When the parties appeared, the magistrates and the jury heard the evidence. Almost anything was acceptable so long as it came from people over fourteen. The court pondered assorted business papers and depositions from witnesses living at a distance or otherwise unable to testify in person. It also heard oral testimony. When the data were in hand, the magistrates defined the law and the jury brought in its verdict. If it went to the plaintiff, the marshal or executive officer of the court proceeded to an "execution" on the defendant's property and turned over to the successful party the equivalent of the sum awarded by the verdict. Costs of court were also allocated, normally to the person losing the action. Appeals from the verdict were open. So was the road of asking for a "review" of the whole proceedings. It was also possible for the loser to "attaint" the jury for error or corruption, and actually to prosecute it in a later action. Finally, in the case of disagreement between judges and jury, or when "special verdicts" reflecting doubt as to proper procedure, or a finding of *non liquet,* or "It is not clear," were brought in, the issues involved could be carried to the General Court for settlement.

Each step of the proceedings was open to challenge by the contending parties, who could now protest that a given attachment had been served illegally, now insist that a certain man ought not to sit as magistrate because of relationship to one of the parties to the suit

---

[10] Richard B. Morris, *Studies in the History of American Law* (New York, 1930), 49–51; Zechariah Chafee, Jr., "Introduction," *Records of the Suffolk County Court, 1671–1680* (2 vols., Boston, 1922), I, xxxviii–xxxix.

or because of private interest in the case, now offer extended written "objections" to the opposition's claims, now fight the issuance of an execution on one ground or another. Poverty was no bar to bringing an action; an act of 1642 made it possible to sue *in forma pauperis*. A marked scarcity of men with real legal training led to no paucity of legal actions; it may have contributed to what must appear to modern eyes as their quite inefficient handling. There were few rules of evidence. Much that was offered was extraneous, irrelevant, and immaterial, and yet, thanks to the provision for appeals and objections, it was heard again and again. True, there was a law against barratry, but it was invoked only well beyond the limits of what a modern court would consider full and proper consideration and adjudication of a particular case.

In the ironworks litigation all of the various objections to various steps in the proceedings mentioned above were used, and more besides. Not only did every conceivable legal tactic or dodge get called into play, but they and their proliferations were heard at every level of the judicial structure. Copies of copies of copies of earlier depositions were left liberally sprinkled through the records of Essex and Suffolk county courts, the Court of Assistants, and the General Court. These materials, quite invaluable to the present study, richly testify to the manner in which the ironworks legal morass imposed terrible strains on a judicial system which was more than adequate to handle most of the cases which came its way.

And, of course, the lawsuits and contentions were not confined to Massachusetts. Issues originating in the New England activities of the Company of Undertakers were also heard in the Sheriff's Court of London, the Court of King's Bench, and Chancery. Some of the cases were, in effect, appeals from the Massachusetts courts by parties who felt that they had been injured by the workings of justice in the Puritan colony. Others were new actions. Both raised all kinds of problems as to proper legal jurisdiction. Uncertainty as to the relationship between the courts of colony and mother country, and the facts of geography, the physical separation, often of contending parties, by the Atlantic Ocean, made their handling awkward, their settlement all but impossible.

It was John Gifford who fired the first gun in what was to become an almost interminable legal battle. In September, 1652, he had come through with his version of the Company's transactions with Captain Tyng, shareholder, supplier to the ironworks, and now a commissioner. Finding it unacceptable, the commissioners demanded a complete accounting. Gifford procrastinated, insisting that he needed to see Tyng's accounts before he could get his own into proper shape. Tyng, however, died unexpectedly, and any chance Gifford may have had to inspect his books was necessarily long postponed. This gave Gifford grounds, however shaky, to insist that it was the commissioners who had failed to come to an accounting, not he, and it also gave him occasion, if one were ever needed by this furious litigant, to go to court.

On June 28, 1653, he sued the commissioners in Salem Court to recover £10,000 which he claimed to have disbursed for the Company. The court referred the case to the commissioners. Two days later the agent petitioned that the court order the commissioners speedily to examine his accounts from the very beginning of his employment. He was requested to hand over his accounts to the commissioners. Tyng's and Webb's were to be made available to him. In the event of disagreement, the court provided, all of the accounts were to be gone over by a group of neutral auditors, the commissioners being allowed a fortnight for their scrutiny, the auditors a full month in which to study the records.[11] Since agreement was out of the question, the case was soon handed over to the auditors. Their task could not have been easy under the best of circumstances. It was, as things worked out, also long postponed. From accounts which had been badly kept, and in which there were gross discordances in the versions presented by the contending parties, they were to determine who owed what to whom—and how well or badly commissioners and agent had served as stewards of the Undertakers' property. Solomon himself would have had trouble!

Their worries doubtless intensified by the clash between the Company's old and new representatives, which must have been the talk of the town in Boston, Lynn, and Salem, the creditors also turned to the courts. Even as the auditors went to work at reconciling the accounts, the creditors petitioned the General Court for a special court to meet at Boston and consider their claims. Their request was granted on August 30, with the significant restriction "provided the

[11] *Essex R. & F.*, I, 284, 286; II, 75.

commissioners for the Undertakers . . . shall agree."[12] Meeting on September 14 and 15, 1653, presumably with the commissioners' approval, the special court heard and found valid the claims of eleven creditors amounting to more than £3600, and granted execution on specified assets of the ironworks in satisfaction of them. The suit was entered in the name of Thomas Savage, a wealthy Boston merchant, to whom, for the sake of convenience, the other creditors had assigned their claims. The defendants were "the Undertakers of the Iron Works, Mr. Bex, Mr. Henry Webb, etc." Webb, however, was also on the other side; at least he had signed over some £350 of his own claims to Savage. This single case saw the greater part of the Company's assets handed over to its creditors.

Gifford, of course, still had claims against the Company, and on September 27 an Ipswich session of Essex County Court heard his side of the story. He sued John Becx and Company to protect himself from still another set of claims, those of the workmen for almost £1365 in unpaid wages. The court found that the Company was responsible for these debts, not Gifford personally, and granted an attachment on what appear to be assets not already covered by the verdict of the special court at Boston. Execution was postponed, however, until December 5, presumably because the auditing, which was expected to clarify the main issues in the ironworks squabble, was not yet complete.

The agent then went to work on the commissioners, and on the creditors to whom most of the Undertakers' property had been awarded. In October, he sought a special court in which pressures might be applied to both parties. Suing Savage and the others for "not giving account of the appraisal of the company's estate, etc.," and Webb, Foote, and Bridges for "putting him out of the company's employment before his time without any reason," he so persuaded the court that it granted an attachment on certain properties at Braintree on October 12. Eight days later, however, the defendants in that case were able to convince a regular session at Salem that Gifford was no "stranger," and therefore not entitled to a special court. While Gifford claimed that he had called the special court to make his enemies show cause why he had been dispossessed of his principals' estate, and to obtain a proper picture for transmission to

[12] Mass. Arch., 59/50, 56; Suffolk R. & F., 225.

the Undertakers, to whom he proposed shortly to repair, Webb insisted that his real purpose had been to get an execution of the September 27 verdict and make off with the proceeds. Actually, Gifford neither went to London, at least for some time, nor vanished as the scoundrel that Webb claimed he was, for the good reason that the commissioners quickly brought him before the bar of justice—at still another court.

The commissioners, if we may believe Webb, fully aware of Gifford's wiles, brought two suits against him in Suffolk County Court at Boston, one in the amount of £25,000 for "breach of covenant," one of £15,000 for failure to give an account of all the iron he had made since first coming on the job. The jury found Gifford guilty of breach of covenant but failed to assess damages pending a complete audit of his accounts by a committee appointed by the court. The pertinent court records have not survived. From Webb's letters to the London group and from Gifford's petitions, however, highly interesting, if wholly conflicting, versions of both the issues and the proceedings may be obtained.

According to Gifford, he had fallen into the marshals' clutches as he went to press his own suit at Salem. He was then attached to Boston court and carried to Boston jail "as a most notorious malefactor." The subsequent legal proceedings involved "such unchristianlike dealings" in Christian New England as few parts of the world could parallel. Charged with failure to provide regular accounts to his principals and their agents, the appropriation of the Company's property to his own use, and risking the ironworks assets in trading ventures of his own, he challenged the legality of the whole business, and insisted that he was being sued solely to prevent his going to England to give his employers a true report of what had transpired. Webb, writing to London, claimed, however, that Gifford had refused to comply with a court order to hand over all his books to the auditing committee. He had announced, indeed, that he would "rather ly in prison 100 yere if he did liue so long." By the agent's own admission he had asked by what law he was being asked to hand over the books which were his livelihood and his only means of clearing himself with his employers. He insisted, though, that he had offered to permit one man to inspect the books at his own house. Presumably the court was in a position to make its order stick, and, with or without Gifford's co-operation, his accounts were eventually turned over to the auditors.

While it looks as though two audits had been ordered, one by Salem Court on June 30, the other by the Boston verdict some months later, it is reasonably clear that the two were somehow merged. As to how this was arranged we are uninformed. Certainly the difficulty of the job in hand would have dictated that one audit was as much, if not considerably more than, the strength of the business experts assigned to the task could bear. To their trials and tribulations we shall return at several points as we try to unravel the ironworks legal maze.

At this stage of the game, however, Gifford and the commissioners were close to stalemate. The agent had lost out at Boston Court but the Essex County verdict covering the workmen's wages still stood. The Company's principal assets had been handed over to the merchant creditors, with the approval, if not the connivance, of the commissioners. Debts to the workmen had therefore to be made good either by the creditors or the commissioners. The former, understandably enough, chose not to pay, and, in mid-December, Gifford obtained execution of the September 27 verdict in his own and the workmen's behalf. This was ordered on the remaining assets of the Undertakers and, should this be insufficient, the personal estate of Joshua Foote and Henry Webb. The claims outweighing the assets, the marshals turned to the commissioners. Foote was out of town and wisely chose to stay there. The blow therefore fell on Webb. While Webb claimed that Gifford and the marshal had broken into his shop and seized his goods, causing, as he put it, "noe lettle trouble to me, & myne, & exceeding great damage to my estate and livelyhoods," Gifford insisted that because of pressures from certain well-placed people, both he and the marshal were clapped into jail.[13] Whatever happened, Webb was clearly almost distraught as he faced full personal liability under the September 27 judgment.

The audit ordered by Boston Court dragged on and on. By the end of February, 1654, it was still unfinished. And by that time Webb was ready to quit. Colonial litigation put heavy strains on all who were engaged in it. Webb had had an enormous dose. He was tired.

---

[13] For details of this phase of the litigation, see the following sources: *Essex R. & F.*, I, 289–95, 309–10; II, 76, 77, 82, 84, 85; Mass. Arch., 59/52, 53, 54; Suffolk R. & F., 225; *Mass. Records,* III, 351; IV–1, 188.

He was so deaf that he had to depend on an assistant to keep abreast of the auditing. He therefore announced his intention to resign his attorneyship for the Undertakers in order "to be freed from these distracted incombrances & to live in peace."

Apart from these personal strains and the liability for large sums of money, Webb's was a highly awkward position. In his dual role of commissioner and creditor he had had much to do with the Company's loss of its property. However eloquently he might describe his strenuous exertions, however black he could paint Gifford's character, however convincingly he wrote that the latter's court victories stemmed from favoritism, there was always the grim fact of the sequestered assets. Once honest enough to admit that, seeing the creditors would swallow everything up, he had decided to join his claims to theirs lest he be ruined himself, he subsequently drew a more circumspect picture. Here, he claimed, he had first tried to persuade the creditors to hold off. Unsuccessful, mainly because they lacked confidence in Gifford and believed that the commissioners lacked the power to fire him, he went along with the plans for a single suit by the creditors, lest a batch of separate actions so dissipate the Company's property that ironworks operations would be impossible, so weaken the ironworks credit situation that the work force would disintegrate. Savage's suit, in other words, was a favor to the Undertakers. The creditors proposed to keep the plants going and keep the workmen together. They would be happy to hand the business back to its owners, once their claims were met.[14]

So far as we can tell, John Becx and his colleagues accepted all this. As late as December, 1654, at any rate, they clearly held John Gifford responsible for their troubles, and this despite the fact that the latter had doubtless been bombarding them with explanations of his own and countercharges to those advanced by Webb. They also accepted Webb's resignation, probably without too much reluctance. By June of 1654 they had replaced Webb, and his fellow commissioners, with a new set of attorneys to John Becx and Company, Robert Keane and Josias Winslow. In the meanwhile, moreover, the General Court had come through with what was in effect a finding of limited liability. It concluded, after no little disagreement, that Webb was not "such an owner or vndertaker of the iron workes as

[14] *Essex R. & F.*, II, 80, 82–83, 85.

makes his person or personall estate lyable to the judgm(en)t of Ipswich Court ag(ains)t the s(ai)d owners or vndertakers."[15] Webb was thus well out of it. Keane's and Winslow's troubles, however, were only beginning.

The new attorneys' first task was to press the suits at Boston Court for breach of covenant and for proper auditing of accounts, and to seek some form of redress from the Ipswich verdict which had gone to Gifford. The Boston proceedings had been carried at least one step toward solution, possibly even before Webb resigned. In March, 1654, the report of the court-appointed auditors charged Gifford with debts of £7979.16s.7d. but allowed him credits of £9631.11s.11d., a finding clearly in his favor. In their report, however, were several pages of controversial items. The case was therefore far from settled. Either by reinstituting the old proceedings or by bringing a new suit against Gifford in the sum of £25,000, with a hearing scheduled for the July session of Suffolk County Court, Keane and Winslow got an attachment slapped on Gifford. Five cows and two or three calves being hardly adequate to meet this huge figure, the hapless ex-agent went to jail on June 16. In the same month the attorneys also petitioned for a review of the Ipswich Court proceedings. The verdict here is nowhere indicated. There is some reason to believe, however, that Keane and Winslow were unsuccessful in their attack from this quarter, and that certain of the smaller assets of the Company remained in Gifford's possession as security for the employees' claims.

In the meanwhile Gifford, too, had been busy. Early in May, he had got off an enormously long and eloquent petition to the General Court. In it he faithfully spelled out his many vicissitudes, raised a number of questions as to the legality of the proceedings by which the Company had been stripped of its property, and begged the Court to appoint a committee, including no Boston resident, no friends of Webb, Savage, and the rest, to go over all the accounts and hear his side of the story. The Court's reactions to the petition are not known. Its reactions to the ironworks litigation in general are somewhat ambiguous—but highly interesting. Two days after Gifford's petition, the deputies were suggesting that a committee be set up to "hear and determine the whole case," on the grounds that it would "require more time than may conveniently be spared from the more proper services

[15] *Ibid.*, I, 347–48, 400; *Mass. Records*, III, 351; IV–1, 188.

of this court." In mid-May, it was of a mind to duck the business altogether; at least the Court voted not to hear the ironworks case, either directly or on appeal from a county court.[16] It had eventually to change its mind, probably for the good reason that Keane's and Winslow's suits offered more grounds for controversy than those brought by their predecessors.

On July 25, 1654, Boston Court ordered the men who had rendered the first audit of the ironworks accounts "fully to examine and perfectly audit" all the accounts between the Undertakers and Gifford. Originally, eight men had been named as auditors, three being designated as substitutes. Six, Nathaniel Duncan, Richard Leader, Thomas Clark, Thomas Lake, Nicholas Davison, and Simon Lynde, now went to work and in mid-August reported a new set of findings. Gifford was here charged with £10,263.2s.4d. of debts and allowed £9725.11s.11d. in credits. Certain problems were still unresolved, however, as, for example, discrepancies between Gifford's and Awbrey's books as to the latter's receipts of iron, and questions as to expenditures for arms, tools for the Scots, clerical help, entertainment, and even Gifford's salary. Passing these over for the moment, it looks as though Gifford was now short nearly £540. He had been nearly £1650 ahead in the first audit. In both cases, however, the sums due the workmen were being carried as Gifford's credits. If he, rather than the commissioners, had to make these good, and the commissioners were strongly of the opinion that he should, his indebtedness under both sets of findings would have been enormous.

Surviving records do not make wholly clear the Suffolk County Court verdict which presumably followed the submission of the auditors' report. There is some reason to believe that there had been disagreement between judges and jury. The latter had questioned some of the auditors' findings. It seems also to have found Gifford not guilty of breach of covenant, or at least refused to assess damages therefor. Judging from two stray documents, the issues were such as to tax the capacities of judges and juries alike. It is also clear that, even though the second audit had tipped against Gifford, the plaintiffs, Keane and Winslow, were still dissatisfied and entered "some just and considerable objections." Whether because of a finding of

<hr />

[16] *Essex R. & F.,* I, 347–48, 398, 401; III, 18; Essex R. & F., III, 18–27; Mass. Arch., 59/52–55, 59; 38–B/181a; *Mass. Records,* IV-1, 194–95.

"non liquet" or because of an appeal by the plaintiffs, the Gifford case reached the General Court in November.

There things went against Gifford from beginning to end. While settlement of the question of breach of covenant was postponed to the following May, the highest court in the Colony accepted the second audit, threw out the first one on the grounds that the men who made it had not concerned themselves with Gifford's following or not following instructions, found lapses, false charges, vast expenses, gifts, and other "improbable disbursements" in his accounts, and overruled his contentions that he had been improperly sued and that two of the judges ought to have been disqualified as partial. The General Court, in other words, seems to have taken Keane's and Winslow's contentions and issued them as legal findings. It wound up by authorizing an execution against Gifford in the sum of £1896.6s.3d., and ordered him to pay the workmen's back wages as well. Quite apart from this last obligation, Gifford was obviously worse off than he had been with either the first or second neutral auditing of his accounts.[17]

The charges against Gifford in these various proceedings are quite plain. The agent had not only been negligent. He was allegedly guilty of embezzlement, misappropriation, graft, and malfeasance as well. The "evidence" which the courts pondered consisted of extracts from the accounts, letters of instruction from London, sworn depositions, orders, and receipts. It was voluminous, and as much was entered by Gifford in his own defense, and in attack on the commissioners and their successors, as was filed against him. Unquestionably, his accounts were badly kept. This alone made it difficult to prove anything against him, even when they were scrutinized by the best business minds in the community. For the rest, it was largely a matter of one man's word against another's. Keane and Winslow might charge that Awbrey had got less iron than Gifford's books showed. The agent could bring up a man who swore that it was Awbrey's books which were deficient, not his. In one case after another the court heard claims that Gifford had beaten the workmen, allowed food to spoil, used the Scotsmen as his personal servants, neglected to take proper care of charcoal, and so on almost ad infinitum. For almost every piece of evidence produced by Webb and Keane and Winslow, there

[17] Suffolk R. & F., 225, 569, 25894; *Essex R. & F.*, I, 401–402; *Mass. Records*, III, 370–72; IV–1, 217–20; Mass. Arch., 59/62.

was usually a directly opposing claim from a Gifford supporter. It is not strange that conflicting verdicts were reached.

To some extent, of course, there was a pattern in the conflicting verdicts. The Essex County Courts were better disposed toward Gifford than those of Suffolk. Henry Webb suggested that his successes at Salem and Ipswich had been due to favoritism. Whether this went back to sympathy toward a neighbor or to a cleavage between the city, Boston, and the more or less rural area to the north is open to question. Similarly it is quite impossible to demonstrate that in the disagreements between magistrates and juries and between upper and lower houses of the General Court, there had been a predilection for what might be called "big business" from the more substantial people of the Colony. Gifford, it is true, expressed at one point his fear that Boston residents might be partial to the prosperous merchants to whom the Company's assets had been awarded. On the other hand Webb insisted that Gifford enjoyed the support of some of the chief figures in Masschusetts, and the Undertakers, at one stage, that Gifford, with their money, had made so many friends that they could not hope to obtain justice. In the case of the General Court, the deputies seem to have been more impressed with Gifford's arguments than were the magistrates. It is highly doubtful, however, if the attitude of the upper chamber reflected special sympathy for the commissioners, for Keane and Winslow, and for the ironworks merchant creditors, rather than concern for the Undertakers' interests in general. Special or external considerations may have been present throughout. The records that have survived hardly permit their isolation, let alone their evaluation.

Rightly or wrongly, Gifford, at the end of 1654, was down. During 1655 the balance shifted slightly to his side. When the action for breach of covenant came up before the General Court in May, Winslow failed to appear, and Keane declined to give security should Gifford recover against him, announcing that he would rather yield his commission than do so. The case was therefore nonsuited. Data are incomplete. It appears, however, that, in the interim, Gifford had sued Keane for false imprisonment and had at least succeeded in making his adversary personally liable should he come out on the wrong end in legal actions in which he was involved as attorney to the Company of Undertakers. While the General Court first deferred hearing the remainder of the Gifford case until the following October, it saw fit to listen to an "importunate request" from the ex-agent, and set the trial for June. At this point Winslow appeared and announced that he, too, would voluntarily relinquish his commission from Becx and Company, doubtless from motives identical with Keane's. Gifford had thus won at least a partial victory in forcing the resignation of his adversaries.[18]

The June session of the General Court began on an ominous note. It was announced that it would hear the *whole* of Gifford's case. Since there were no defendants to the ex-agent's charges in a bill of review, the business was put over until October. The Court freed Gifford from jail under £2300 bail, though on condition that he be confined one day and night per month, and granted him the right to copy from his "broad book" over an eight-week period in order to build up his case against whoever might then turn up to oppose him. In October, the Court decided that Gifford's oath was not sufficient to validate his accounts, and that his "proof" of Awbrey's receipts of iron was unacceptable. To reach even these conclusions took several days, and the case was still unsettled. Inevitably, there was still another postponement, and, again, Gifford was free to go about on bail, presumably gathering additional evidence for his various contentions.

In mid-November, at long last, came at least a partial settlement of the complex issues. After much pondering, and the resolution of disagreements between magistrates and deputies, the Court saw cause to reduce the amount of Gifford's liability from the £1896.6s.11d. of the original verdict to £1225.19s.1d. At the same time, it announced formally that Gifford had not been able to justify various disbursements, and reiterated its decision that he was responsible for the workmen's wages.[19] Thus, though somewhat better off, the ironworks manager was still much in debt. Inevitably, he went back to Boston jail. He may have been able to frighten off his adversaries; he had not succeeded in bringing around the magistrates of Massachusetts to his side.

To this point we have been handling the question of liability for the debts to the workmen as incidental to the main struggle between

[18] Mass. Arch., 59/54, 59–61, 64, 65, 65a; *Essex R. & F.,* I, 398–402; II, 84; III, 18–27; *Mass. Records,* IV–1, 228; 237, 241.

[19] *Mass. Records,* IV–1, 241–44, 251–53.

Gifford and the commissioners, and their successors. Actually, of course, the workmen's claims, and those of a number of small creditors in Essex County, were both real and pressing. When the ironworks business troubles reached their climax, no one was as badly off as the workers at Hammersmith and Braintree, and neighboring farmers and small shopkeepers. The former were utterly dependent on their wages, the latter could ill afford to have payment of even small sums long delayed.[20]

Liability for these obligations was charged, at various points along the line, against Gifford, the commissioners, and the large creditors to whom the bulk of the ironworks property had gone in the fall of 1653. The commissioners insisted that the manager was properly responsible, since the debts had been incurred in the course of normal business operations, and should have been covered by sales of iron. Gifford held the commissioners liable. They were the owners' representatives and the debts were in the Company's name. The large creditors' responsibility for debts contracted long before they entered the scene is not as ridiculous as it might appear. There is some reason to believe that their being awarded most of the plant had been conditional on their undertaking to make good on the smaller claims.[21] They never did so, however, and it was therefore between Gifford and the commissioners as to who should pay.

When the General Court settled the question by finding Gifford responsible in November of 1654 and 1655, the judgment and its reiteration brought an approximation of order out of a chaos of litigation almost as extensive as that to which the central issues in the Gifford case had given rise. Between the fall of 1653 and the spring of 1656 more than twenty-five suits to recover on small debts of the ironworks were filed in the courts of Essex County.[22] In the early actions, there was much confusion as to whom to sue. Most of them were left hanging, presumably until the larger ironworks problems got resolved. This was no help to the small creditors. We have seen that Ipswich Court awarded Gifford a share of the Company's assets from which to satisfy such claims. To these he held on, at least until the November, 1654, General Court verdict. In May of that year,

[20] Mass. Arch., 59/54.
[21] *Essex R. & F.*, II, 79; Mass. Arch., 59/54.
[22] *Ibid.*, I, 300, 332, 372, 374, 378, 386, 393–94, 417.

Joseph Armitage and "divers other poor workmen" called the magistrates' attention to the plight of needy families "craving supplies" and "in deep extremity," and begged them to order Gifford to pay them out of his Ipswich execution, or the commissioners to make good their urgent claims out of the Company's estate.[23] Action on this petition was, presumably, also postponed. When the General Court issued its first judgment against Gifford, there was a flurry of suits in Essex County Court, all of them filed, properly enough, against the Company and Gifford. Thirteen plaintiffs emerged victorious and owning ironworks land and other property which Gifford had presumably been holding.[24] There were a number of similar suits during 1655, but there is no record of transfers to the small creditors who brought them, probably for the reason that there were no more known ironworks assets to distribute. Those to whom money was still owing could collect only from Gifford himself.

That gentleman was presumably penniless and certainly still confined to jail. In the spring of 1656, however, he presented to the General Court a letter from John Becx and four of his fellow Undertakers, in which the authorities were asked to release the former agent from jail and allow him to proceed to England with all his books and papers.[25] The tone of the letter was neutral, and it is impossible to tell whether Gifford had finally persuaded the Undertakers to hear his side, or they had concluded that to do so was their sole remaining hope. The Court's reaction to the letter was favorable, and on May 14 it set Gifford free, insisting that he discharge his prison fees but saying nothing of his other obligations.[26] The costs of incarceration for more than three years somehow met, probably by friends, Gifford set sail for the homeland.

Reaching England in October, he went to London and settled down to long discussions with his principals, and an auditing of his accounts. The latter job was handled at a Fleet Street tavern over a two-week period early in 1657.[27] In it all of Gifford's contentions were borne out and the ex-agent emerged, not only cleared, but with a power of attorney from the Company with which to begin a new round of

[23] Mass. Arch., 59/58.
[24] *Suffolk Deeds,* II, 265–72.
[25] Mass. Arch., 59/71.
[26] *Mass. Records,* IV–1, 268.
[27] Add. MSS. (British Museum), 34015, 70; C9/21/5.

skirmishes against the creditors who now held all of the ironworks assets! It is difficult to ascribe a complete reversal of the manager's role to eloquence alone. His accounts must have proven acceptable to Becx, and Becx must have been better qualified to evaluate the ironworks bookkeeping than those who had wrestled with Gifford's records in New England. Indeed, one might be inclined to take this as prima facie evidence of the ex-agent's innocence. The fact that all of the Undertakers were not equally convinced and eventually brought suit against Gifford, one of his clerks, *and* John Becx, charging conspiracy against the Company, rules out too quick a judgment on this point.

An ostensibly all-victorious John Gifford returned to Massachusetts with power of attorney in hand and fond hopes of revenge in his heart, late in 1657 or early in 1658. In May, he petitioned the General Court for a review of the legal proceedings from which the large creditors emerged as owners of the ironworks, suing Thomas Savage for "withholding the estate" of Becx and Company, and asking for damages. Through five days of hearings the two sides presented their evidence, the atmosphere doubtless as tense as the legal crisis was important to both Massachusetts and the contending parties. Then the Court announced its verdict. While it admitted several "circumstantial errors" in the old Special Court trial, it found no reason, "according to aequitje and justice," to reverse the earlier judgment.[28] Gifford had lost again.

In June, he was back at his old haunts, the courtrooms of Essex County. There he sued Henry Webb, his late chief adversary, for "defaming the plaintiff to his principals in England by writing, rigorous handling, vexing, prosecuting, unjustly molesting and imprisoning him," and for "unjust molestation."[29] In the court records, these are carried as two separate cases. In the first one, the principal piece of evidence was a June, 1653, deposition charging Webb with enough lapses to suggest outright culpability in his dealings as a commissioner with Gifford. For the second, a welter of data got presented. There were Webb's old letters to the Undertakers, which Gifford must have obtained from John Becx, and which, as we saw earlier, contained much that was, at very least, quite hard to explain. To back up what

[28] *Mass. Records,* IV–1, 330–31.
[29] *Essex R. & F.,* II, 71, 74–79.

he had written, Webb introduced witnesses and sworn statements which charged Gifford with one sin of omission or commission after another. In effect, all the old issues were reheard. This time it was the plaintiff, Gifford, who came out on top, and in both cases.

Was all this just another instance of Essex County partiality to Gifford? Was it the consequence of his having now appeared with the patent endorsement of some of the very people in whose eyes Webb had tried to blacken his reputation? Again, it is hard to tell. We know that Webb tried to get these actions quashed. He challenged the competence of one of the judges. He questioned the propriety of bringing the suits in Essex County, where neither plaintiff nor defendant then resided. He claimed that the whole proceedings were *ex post facto,* since he had long since renounced his attorneyship, and the main issues had long ago been settled by the General Court, where alone the present case was triable, if at all. His objections or his evidence, or both, had no influence on the jury. They did persuade the magistrates to withhold their assent to the jury award of £500 to Gifford.[30] Probably because of the disagreement between judges and jury the matter went to the General Court in the fall of 1658. The Essex Court data were handed over to William Hathorne, the man to whose presence as judge Webb had objected, for transmission to the higher court. There the issues were apparently pondered in mid-October. By that time, however, Gifford was back in England again, and Hathorne was looking out for his interests and pleading his case under a power of attorney.

For all we can learn from records which are strangely silent on the matter, the General Court "settled" this new version of the Gifford case by finding his power of attorney to Hathorne "not good in lawe."[31] The basis of this finding is nowhere indicated; data on the handling of the real meat of the litigation are also lacking. One wonders if the absence of pertinent material in surviving records is wholly accidental, if the repudiation of the power of attorney had been a handy means of escape from a dilemma of the Massachusetts judicial system. The General Court had found against Gifford, time and again. Now he had not only the support of the Undertakers but at least a partial vindication from Essex County Court. His cause,

[30] *Ibid.,* II, 413; IV, 30.
[31] *Mass. Records,* IV–1, 352.

or, more properly, the cause of the Company, had also received influential backing from a new and most important quarter, from Oliver Cromwell himself.

In June, of 1658, even as Gifford was preparing to bring his actions against Webb, the General Court found itself in an extremely ticklish situation. It had to draw up "some meet and substantial Ground and Reasons of the Court Judgment" in the case of Gifford, as attorney to Becx and Company, versus Savage, "to be sent to England." The product of a special committee's efforts, a draft of a letter which makes it plain that the Lord Protector was its addressee, has survived. This fascinating document somewhat piously announces that the General Court had heard all the charges, pondered all the evidence —and come up with a decision with which no impartial hearer could quibble, least of all Cromwell, in whose "favor & justice" the Court had full confidence.[32] Just how the Court received the Protector's "pleasure" in the first place, and how he reacted to the magistrates' letter, are presently unanswerable. With little question, it had been the channels that Becx used in taking on his Scots prisoners that got Cromwell interested in the Company's difficulties. The worries of the gentlemen who presided over the destinies of Massachusetts as the ironworks litigation threatened to spill over into the troubled waters of relations between the Puritan Colony and the Puritan Commonwealth may readily be imagined.

From October of 1658 until April of 1662, Gifford remained in England, indulging to the full the passion for litigation with which he had got almost nowhere in Massachusetts. His first victim, as might have been expected, was Henry Webb. In April, 1659, he persuaded the Sheriff's Court of London to attach a hogshead of beaver skins belonging to Webb, for an alleged debt of £300. Over the protests of Webb's attorneys he was awarded a judgment of £50. The attorneys appealed to Chancery, and in proceedings which involved, among other things, a scrutiny of Gifford's 129-page, parchment-bound account book, Webb, in 1662, won his case and was awarded costs.[33] In the interim, after developments of which we have no record

[32] Mass. Arch., 59/80; Suffolk R. & F., 290.

[33] Historical Manuscripts Commission, *Seventh Report* (London, 1879), 87; Suffolk R. & F., 469; C9/21/5; C24/871; C38/149. Gifford seems not to have seen fit, perhaps for good reason, to leave the accounts in court, where they might have been preserved to this day.

but which must have been most interesting, Gifford had turned against his former employers. In an action in the Court of King's Bench, he sued Becx, Bond, and Pocock to recover his salary. Here he was successful and was awarded £400 and £26 costs in Hilary term, 1660.[34] The money must have been most welcome, if only because Gifford loved lawsuits, and actions at law were expensive, particularly for those who lost.

The Restoration found the ex-agent petitioning the King concerning the government of Massachusetts. His checkered career in New England made him an eminently logical first signer among a group of people who had suffered in one way or another at the hands of the Puritan fathers at Boston. Here he and the others recited, among other things, that:

Through the tyranny and oppression of those in power there, multitudes of the King's subjects have been most unjustly and grievously oppressed contrary to their own laws and the laws of England, imprisoned, fined, fettered, whipt and further punished by cutting of their ears, branding the face, their estates seized and themselves banished the country. . . .[35]

Sentiments like these, strong as they may have been, did not prevent Gifford's return to the Bay Colony, as we shall shortly see. One must assume, however, that only the magistrates' ignorance of this petition lay back of his being allowed to disembark on Massachusetts soil.

In still another series of proceedings, Gifford was on the receiving end. Apparently in retaliation for the suit to recover his salary, Bond and Pocock brought an action against Gifford, in Chancery, in which Becx and one John Blano, a former clerk of Gifford's, were named as codefendants. This case dragged on, with interruptions, until 1663. The plaintiffs charged Gifford with gross dereliction of duty, and him and the others with conspiracy to take over the Company's property. Despite the fact that almost all of the ironworks assets were by this time in the hands of the Company's Massachusetts creditors, they insisted that the defendants had all or most of the shareholders' estate in their own custody, or had "disposed of it and shared the proceeds." After still another survey of Gifford's accounts, and hear-

[34] C41/16, Easter 460.
[35] C. S. P. Col., America and West Indies, 1661–68, pp. 16–17.

ings in which it was demonstrated to the Court's satisfaction that Becx and Blano had been named as defendants in order to stop them from giving evidence in his defense, he was eventually acquitted.[36] Even before the last gun in the battle of Bond and Pocock *v.* Gifford had been fired, however, the ardent, and seemingly at last really successful, arch-litigant had set sail for Massachusetts in April, 1662.

The Chancery verdict is by no means proof perfect of Gifford's innocence; after all, the General Court had come to a directly opposite conclusion. The case just summarized, however, produced no evidence of collusion to a court which was specifically looking for it. This, and the fact that Gifford had coupled Becx's name with those of Bond and Pocock when he sued for his salary in the Court of King's Bench, suggests that, at very least, we put behind us all doubt as to the authenticity of the stamp of approval that Becx had placed on the manager's accounts soon after he reached England. Circumstantial evidence seems also to tip in Gifford's favor. No evidence of a concealed estate ever came to light, although one must wonder where a man who had hardly been well off when he entered the Company's employ got the money with which to fight his innumerable lawsuits. Certainly, however, it was with borrowed money that he financed most of his later American activities.

The Massachusetts courts had not been wholly free of ironworks litigation during Gifford's absence. There had been suits over old debts, in which now Webb was under attack, now Hathorne was trying to recover as attorney to Becx and Company, and again the Company, presumably with Hathorne as attorney, defending itself in a case whose issues ran all the way back to Leader's time.[37] With Gifford's return, however, legal activity was stepped up considerably. In June, 1662, he sued the "executor, administrator or possessor" of the estate of Robert Keane for the the latter's having taken away Gifford's goods and wrongfully imprisoning him. The case was thrown out because, despite all his experience with the law, he had named the wrong person as executor.[38] In the same month he reopened the old case against Webb by petitioning Essex County Court for an

---

[36] C9/21/5; C9/24/9; C9/22/24; C22/256/7; C24/857; C33; C38; C41; C33/217, fol. 264; C38/146.
[37] *Essex R. & F.,* II, 127–28, 184, 193.
[38] *Ibid.,* II, 389.

execution of his £500 judgment, conveniently forgetting the General Court's disapproval of his power of attorney to Hathorne, and alleging that it had left the case unsettled. Here the lower court took the quite logical step of referring the matter to higher authority.[39] Not long thereafter he had to defend himself in a suit lodged by Margaret Sheafe, Webb's daughter and executrix, to recover the £50 he had won at the Sheriff's Court of London. Started, apparently, in Suffolk County Court and appealed to the Court of Assistants, proceedings which involved everything from challenging the validity of Gifford's claims against Webb to questioning the due and proper administration of old-country justice guaranteed that the problems be pondered, sooner or later, by the General Court.

In October of 1652 the highest tribunal deliberated both the Keane and Webb cases. Its initial impulse was, once more, to duck. A committee announced, at any rate, that the issues ought to be thrashed out in the courts in which they had earlier been heard. The whole Court, however, soon came to a different conclusion. The Keane case was thrown out on the grounds that Gifford had not brought suit against him during his lifetime. In the Webb business, the Court pointed out that neither Gifford nor his attorneys had brought new actions after Hathorne's power of attorney had been ruled invalid. Of both, it announced that there was "no ground on these verdicts to proceede to judgment, the defendants being deade."[40] In the eyes of the Puritan fathers, in other words, death was enough to put people beyond the reach of Gifford's clutching legal arm.

By the end of 1662, then, Gifford had clearly lost out on both his main fronts. His efforts, both to recover the ironworks from their new owners and to wreak vengeance on old enemies, dead or alive, had come to naught. Perhaps duly chastened, perhaps too occupied with private concerns, he put ironworks litigation that went back to his Hammersmith connection behind him for more than six years. In June, 1669, he brought, and lost, a suit to recover on an old debt of the Company of Undertakers. In March of the following year he successfully defended Becx and Company in suits filed by Joseph Armitage as assignee of two of its old employees, Samuel Bennett and Richard Post.[41] Armitage, every bit as litigious as Gifford, was no

---

[39] *Ibid.*, II, 413.
[40] Suffolk R. & F., 469, 485, 568; *Mass. Records*, IV-2, 65.

mean adversary and would not stay down. He prosecuted the old Bennett claims in June and November of 1670, and in March, June, and November of 1671, losing out, apparently, in all but the last. There, however, he won his case, and got an execution on certain parcels of land which had somehow remained in the possession of Becx and Company despite the various forays of both small and large creditors.[42]

In the interim, Gifford had been having another go as plaintiff, this time on an old ironworks claim against George Corwin. Successful in June, 1671, he lost out in September. When the case came up again in November, a special verdict was forthcoming, one which was neither victory nor defeat for Gifford. Armitage's success and the vagueness of the special verdict were enough to persuade the embattled ex-agent to call it quits. In the same month, more than twenty years after he first went to work for the Undertakers, he renounced his attorneyship to Becx and Company.[43] There must have been many who sighed with relief!

Gifford had not been the only man bent on recovering the Company's assets from their new owners by legal action, however. It is not at all unlikely that the controversy between Bond and Pocock on the one hand, and Gifford, Becx, and Blano on the other, had split the Company into factions. When Gifford left London in the spring of 1662, there was an affidavit on file in an English court in which Pocock and Hiccocks, two of the partners, deposed that "The Undertakers . . . doe imploy others . . . to be their Commissioners to receive their works." By inference from documents which survive in Massachusetts records, the "others" were Joseph Hill and Richard Collicott. In the fall of 1662 the General Court turned down a petition which they had presented "in the name & behalfe of the adventurers & copartners of the iron works." In all likelihood, they were asking for a review of the proceedings that had awarded the Company's property to its creditors in 1653. The following spring they brought suit in Essex County Court against Thomas Savage, the merchant to whom, legally speaking, the assets had gone. Suing as attorneys to "Mr. William Becks, Lionell Copley and others, the adventurers

---

[41] *Essex R. & F.,* IV, 150, 220–21.
[42] *Ibid.,* IV, 262–63, 298, 339, 383, 431–32, 451.
[43] *Ibid.,* IV, 391, 421, 436, 445.

and co-partners in the Iron works of Linn and Brantry, formerly known and called by the name of Mr. John Bex & Co.," they were wholly unsuccessful. The court found their power of attorney invalid on the grounds that it was signed by only six partners whose total holdings they could not demonstrate to outweigh those of Becx and others whose names did not appear on the instrument. The General Court subsequently upheld this finding.[44] Hill's and Collicott's efforts were thus no more fruitful than Gifford's had been. One must wonder, however, if the Massachusetts courts had not had more than enough ironworks problems to wrestle with without being asked to decide who properly represented the Undertakers.

Only the later developments in that secondary but not unimportant area of ironworks litigation, the claims of the workers and other small creditors, remain to be considered. These were, in the spring of 1656, when we left them to follow John Gifford in the main stream of lawsuits and contentions, far from settled. We have seen that certain workmen had acquired title to certain low-value assets in 1654, that a number who sought to do likewise in 1655 seem to have fared less well. In June of 1656, however, five suits were filed in Essex County against Becx and Company and John Gifford, with all but one carrying the designation "for work done at the Ironworks." Various minor creditors were apparently bent on getting a last crack at the Company before its late agent "escaped" to England. While the verdicts are not indicated, it appears as if some transfer of assets took place at this time. Joseph Armitage, for example, acquired an interest in the slitting mill and a share in a conduit in Boston in one or another of three actions in which he had been plaintiff. Similarly, he, as assignee to Samuel Bennett, and Daniel Salmon as administrator of the estate of Joseph Boueye, a part-time worker, got three hundred acres of Braintree land in satisfaction of £110 worth of claims. There was another side to this coin, however. Even as the workmen sought to recover from Gifford, that gentleman, through William Hathorne and Amos Richardson as his attorneys, was suing in order to recover on debts *from* some of them.[45] Again, the verdicts

[44] C41/15, Hilary, 1662/3, no. 76; *Mass. Records,* IV–2, 61, 85; *Essex R. & F.,* III, 41–42.
[45] *Essex R. & F.,* I, 424–26; Essex Deeds, I, 33; *Suffolk Deeds,* III, 3, 30–32.

are lacking. Presumably, however, the people in question were in poor position to make good.

It is difficult to escape the conclusion that many of the small claims against the ironworks were never satisfied. Broadly speaking, the cards were stacked against the men who held those claims. Procedure in the midst of a gigantic legal morass would have come hard for even the better informed among them. The courts, with the best intentions in the world, would have had trouble adjudicating small parts of an enormously complex whole. And above all, of course, the larger creditors had got there first. Whatever Gifford may have been handed in his Ipswich Court execution, it could not have been enough to go around. Some got some satisfaction from the courts. Some may have been paid off by the Company's agent or commissioners. With the passing of the years many of the smaller claims must have been forgotten, even by those whose need had once been greatest.

Inevitably, one wonders to what purpose had this all but endless litigation been. The Company of Undertakers had lost its ironworks, its land, essentially everything it owned in New England. The questions of individual culpability, which had been so thoroughly argued in so many courts in the homeland and in Massachusetts, are still questions. Gifford may have been a scoundrel; so may Awbrey, Webb, and some of the others. Was justice done? No man can now tell. If so much remains in comparative obscurity, one thing at least is plain. The long and intricate record here summarized surely makes it clear why the "contention and lawsuits" overshadowed everything else in the eyes of old Hubbard. His judgment was sound. Fortunately, however, iron was in fact being made at Hammersmith and at Braintree, even as the endless suits and countersuits were being carried. All appearances to the contrary notwithstanding, productive activity did not cease with the eclipse of the Company of Undertakers in the fall of 1653, nor for a long time thereafter.

12

Colonial Proprietors Try
—and Fail

IT IS DOUBTFUL if any man in Massachusetts expected the sequestration of the assets of the Company of Undertakers in mid-September of 1653 to be permanent. The proceedings culminating in the Boston Special Court verdict had been initiated with reluctance. The ironworks' creditors claimed to have exhausted all other possibilities before taking their claims to the courts. The Company's commissioners, their right to mortgage property open to question and their efforts to pay off old debts by borrowing new money all but forlorn, saw no alternative to letting the law take its course. The magistrates were the most reluctant of all, probably because of a sense of responsibility to both sets of parties to the controversy. Having to protect the interests of the English capitalists with whom they had co-operated to bring ironmaking to Massachusetts, but wanting also to look out for their own people, they were caught up in a dilemma which the prominence of various of the individuals involved can only have made singularly pressing.

Before authorizing the calling of a special court, the General Court had unsuccessfully urged a "compliance," or interim working compromise, on the Company's creditors and commissioners. It had pondered petitions from both sides as to how to handle the bills of exchange which the Undertakers had refused to honor. It had duly deliberated the question of whether or not the commissioners' power of attorney covered a so drastic action as the mortgaging of Company assets. When, finally, it passed an order for a special court to resolve the ironworks financial mess, it demanded and got the commissioners' prior agreement,[1] doubtless out of tender concern toward the English shareholders in the Company.

Significantly, too, it was a special court to which the pressing ironworks problems got handed, a court which was, by legal definition, a merchants' or strangers' court set up to administer justice on at least something of an emergency basis. Only the emergency could have excused the seeming irregularity of Captain Savage's being allowed, "for the more expedite dispatch of the cause," to bring an action of debt in his own name, and as assignee to ten other creditors, in the sum of more than £3600.[2] Only an expectation that the awarding of most of the ironworks plant to the creditors, or, strictly speaking, to Savage himself, was but a temporary stopgap solution to the ironworks business difficulties could have justified the concurrence in the Special Court's verdict by magistrates, commissioners, and creditors alike.

Practically speaking, of course, the property went to a group of people, the joint creditors. The legal basis on which their claims were pooled is not spelled out in surviving documents. The propriety of the action was subsequently questioned. Similarly, the formal basis on which the group proposed to work the plants is nowhere specified. It is unlikely that more than an informal working agreement was involved, since just as clearly as the creditors proposed to keep the plants going, so they hoped that they would have to do so only for a limited period of time. They expected, and the Court itself may well have expected, that the Undertakers would quickly retrieve their properties by composition with the creditors.

Early in November, 1653, the creditors got off a letter to John Becx and Company which summarized the background of the Special Court proceedings, essentially as outlined above. Their reluctance in the whole business was duly stressed. They indicated, indeed, that in turning to the law they had been acting as much in the Company's

---

[1] Ironworks Papers, 261; *Essex R. & F.*, II, 79; Mass. Arch. 59/56, 150; Suffolk R. & F., 225. It is interesting to note that even the creditors questioned the commissioners' right to mortgage the Company's property.

[2] Mass. Arch., 59/146, 150; Ironworks Papers, 163–66.

interest as in their own, driven to it in no small measure by their awareness of the "improvidence and bad husbandry" of some of the Undertakers' employees. Now, of course, the works were theirs and things were much improved. As soon as they took possession they had provided clothing and supplies for the workmen, put through repairs, and got both furnace and forge into production after a shutdown which they attributed to want of water. The Undertakers had only to make good on their old obligations by the following March, and reimburse the new proprietors for what they might be out of pocket as they operated the plant in the interim, to be restored to possession of ironmaking facilities that would be in far better shape than when they had been handed over to their Massachusetts creditors.[3] Clearly, this was what they wanted.

One would give much to know what was in the mind of John Becx as he received this letter so full of pious disclaimers of any inclination to hurt the English owners, so eloquent in its attempt to make the creditors appear, not as vultures who had seized their property, but as their well-disposed stewards. It did not have its intended effect. We have seen that the Undertakers chose eventually to fight for the return of their estates at the bar of justice rather than to redeem it by paying off their creditors. It is at least possible that the transparent hypocrisy with which the letter seems to reek had something to do with their decision to do battle for their property in every quarter which was open to them.

The creditors made another try in a letter of September 16, 1654. It has not survived. Its contents may be reconstructed, however, from a reply to it by John Becx, which is in many ways as fascinating as the creditors' earlier letter.[4] Perhaps because a year's efforts had persuaded them that running an ironworks was no easy matter, the creditors seem to have offered Becx and Company a more generous proposition, one which would allow them to take back their property at once and to enjoy a three-year period in which to make good on the creditors' old and new claims, with interest, out of the proceeds of the ironworks production.

Becx met this new offer with a recital of various difficulties of his own, and with a counteroffer. He had tried to bring the other Under-

[3] Suffolk R. & F., 225.
[4] Ironworks Papers, 228–36.

takers together to consider the creditors' proposal but had been unsuccessful. This, he suggested, was enough to indicate that his fellow investors had put the New England ironworks behind them. Since, on the other hand, he held not only a quarter share of the business but most of the English claims against the Company, some £2100 out of some £3300, he felt impelled to try to go it alone—although on quite different terms from those which the creditors were now suggesting.

The Becx propositon called for joint operation of the ironworks and a sharing of their proceeds between old and new proprietorships and between Massachusetts and English creditors. Passing over the question of the validity of the new owners' title, he suggested that they, reduced in number from eleven to four or six, stock and operate the plant, hopefully without salary, perhaps with the assistance of a clerk paid £20 or £30 a year, ideally with William Osborne as "vper Clarcke." The profits would then be split, 30 per cent going to pay off the old claims of Massachusetts creditors, 30 per cent, those still outstanding in England, 20 per cent as return on the operating capital being advanced by the new "owners," and 20 per cent on the original investment of the Undertakers.

Becx, obviously, would have come off well in such an arrangement. His own share of the proceeds, as large investor and heavy creditor, would have been impressive. Conversely, because he was already so much in arrears, he declined to put up his half of the stock and supplies until all the English debts had been cleared. The new creditor-owners or, more properly, creditor-operators ought by no means to complain, he insisted, since they were being offered a chance to come in as equal partners "for nothing," to use his phrase, in a business venture into which the original group had poured more than £15,000. Eventually, Becx wrote, they might come in "as purchasers" or part-owners, hopefully by buying up the claims of a few old partners who might still be interested, for a price which he reckoned at less than £1000, and which he indicated might be met in small annual installments. Unless the ironworks really prospered, however, even this might be unnecessary.

If the creditors had been eloquent in depicting their "kindness" to the Undertakers in their Special Court proceedings, Becx here made out a good case for his proposal as a fine thing for Savage and his fellows. He pointed out the advantages of having active partners on both sides of the Atlantic. He dangled before their eyes hopes of profits of at least £1200 a year; indeed, he announced that if he were running the works, he could make £2000 as easily as he could turn a hand. Most important of all, his scheme promised freedom from "all Clamors"; without it the creditors would be plagued with law-suits constantly and interminably. Since, stripped of its eloquence, the whole proposition boiled down to one great big question mark in the crucial area of iron-works *ownership,* however, it would have been strange indeed if it had found favor with Savage and the other creditors. Much as they may have wanted to unload the ironworks, they would have been insane to accept Becx's less than generous, and perhaps less than honest, offer.

Ironworks ownership was, of course, thoroughly snarled even on this side of the Atlantic. Savage had not got all of the Undertakers' property. We have seen that Gifford held a portion of the assets for some time, at least, and that, ultimately, various pieces of property went to various smaller creditors. Given duplications in inventories, attachments and executions and blanks in surviving records, it is all but impossible to tell who owned what at what time. Most, if not all, titles were open to question. Ownership was thus both fractional-ized and confused. Add this to the vague basis on which the eleven large creditors holding the bulk of the property proposed to work the plants, and one must wonder if the glowing reports in their first letter to Becx had much foundation. They got the works in Septem-ber. They claimed that they had Hammersmith in full operation in the first week of November. If so, the new proprietors must have solved a number of problems in very short order.

True, the plant at Lynn was theirs, and reasonably intact. They had the wherewithal with which to stock it and keep it going. Most of the workmen had doubtless stayed on. Since no one of the new proprietors had an iron-trades background, however, the problem of management was especially pressing. This had arisen, of course, as soon as John Gifford had fallen victim to the commissioners' at-tacks. Those gentlemen had come up with no more than a temporary solution. They turned to the obvious candidate, William Osborne, but he declined on the grounds that the business was so much in debt and in so much confusion. They then sought out Thaddeus Riddan,

Leader's old clerk. He had the great merit, in the commissioners' eyes, of being an honest man and one who, in obvious contrast with Gifford, would "liue answerable to his place & incomes, and not lavesh it out, as others haue done."[5] On the other hand it is doubtful if he were equipped to take on the higher-level managerial functions, particularly since he had been away from ironmaking for some time. Well qualified or otherwise, Riddan was probably in charge during the last stages of operation under the commissioners, and certainly so during the first part of the Savage regime.[6]

In relatively short order the new owners provided themselves with an "agent," Edward Hutchinson, a close relative of one of the group. Since he was a Boston merchant, and seems to have remained so during his two-and-a-half-year term as agent, however, it is likely that the on-the-spot management was provided by clerks, Riddan at first, Oliver Purchas from 1655 on.[7] So far as we can tell, the qualifications of neither of these even approximated Gifford's. It is probable, therefore, that the Savage group had been less than wholly successful in filling a spot which was, to judge from the whole experience of the Undertakers, utterly crucial.

There was, too, the need for repairs at the plant. The very testimony which had damned Gifford suggests that these must have been extensive and costly. Similarly, to judge from testimony available in a number of quarters, the feeding and clothing of the workmen, those innocent victims of the business collapse, until such time as they were earning wages again would have been no light burden to the new management. True, Savage and the others had assured Becx in their first letter that all this had been accomplished. One may fairly ask, however, if it had been only the want of water in a season of drought that had plagued Savage as he set about getting the ironworks back into production.

Even if this had been accomplished as quickly as the new owners suggested, it is all but certain that they failed to achieve successful operation of the ironworks. If the works had been booming, John Winthrop would probably not have got a welter of letters from Hammersmith workmen, presumably in the winter of 1654–1655,

[5] *Essex R. & F.,* II, 86.
[6] Essex R. & F., XLII, 136–5.
[7] *Essex R. & F.,* VIII, 197, 404; Mass. Arch., 59/159, 173.

which indicated great eagerness to join him at his new ironworks in Connecticut. More telling still, the General Court would not have broken its ironmaking monopoly grant to the Undertakers, and their successors, in the fall of 1657, when it authorized the setting up of ironworks at Concord.[8]

The Court's answer to the Concord petitioners must stand as the best general picture of the Saugus River plant during the first phase of operations under colonial ownership. The magistrates announced that they had been "credibly informed that the works in present being are not like long to continue," that the works were not meeting the region's need for iron, for the good reason that the present owners sometimes "had it not," and that when they did, they were selling it at more than the court-appointed ceiling price. Even more damaging, the new owners had declined to answer a query from the magistrates as to "whither they held themselves engaged to make good the vndertakers couenant as to the supplying the countrje with iron." Here is evidence enough that Hammersmith was in the doldrums, operating at least fitfully, but far from thriving.

A similar picture emerges if we turn to the record of how the new owners went about handling the properties which the Special Court at Boston had handed over to them. We may assume that their first act had been to divide up moveable assets, like the bar iron, according to their respective claims. They then appear to have settled down to a joint operation of the works during what was initially regarded as the interval between writing to, and hearing from, the Undertakers. Savage gave his "consent" to the others' carrying "a part in the said Iron works" for just such a period. In so doing, he insisted, reasonably enough, that he not be left in a position of sole personal liability should lawsuits by the old Company go against him. It was thus a group of men who advanced supplies to a plant which was "out of credit, victuals and clothes."[9] When no good word came from England, the creditors continued jointly to operate the ironworks.

The structure of the new ownership group is obscure today. It was, even to its employees, in the seventeenth century. Of Savage's centrality in the whole business, there is little question. On at least one occasion, the new group was referred to as "Savage and Company." We

[8] *Mass. Records*, IV–1, 311–12.
[9] Mass. Arch., 59/159.

shall see that he bought up a number of the scattered assets of the Undertakers. He held strict legal title to the whole business and, so far as we can tell, relinquished none of his rights. It was he and Hutchinson who hired Purchas. It is true that the new clerk was ordered by Hutchinson and Webb to trade with Savage "as with another man, noe Owner." Presumably, however, this meant only that Savage had already got his share of the ironworks proceeds, and the time had come for the others, who had held three quarters of the claims against the old Company, to receive some of the products, particularly those men who had seen fit to join Savage in putting up new operating capital. Chief among these were Henry Webb, Anthony Stoddard, and Jacob Sheafe. Hutchinson may also have qualified as a partner, perhaps via his family interest. One stray document, at any rate, refers to the new proprietors as "Mr. Edw. Hutchinson of Boston and Company now of Iron works in New England," a designation which his agent's status alone would hardly have justified.[10]

Data on production and sales under this group's aegis are scanty in the extreme. Between the summer of 1654 and the autumn of 1657, we find only stray clues indicating business activity, and in all cases more than a little obliquely. Evidence of production in the spring of 1658 is somewhat more solid. In June of that year, there arose a legal controversy involving the exchange, at Hammersmith, of a quantity of malt for "bar iron slit into nail rods." The exchange was not accomplished, apparently because of a shutdown of the slitting mill. That bar iron was on hand is clear from the fact that Purchas and two of his workmen got into trouble for violently seizing some which a marshal's deputy was in process of attaching.[11]

On prices we are even worse off. In March of 1657, when a court verdict went against Hutchinson for "unjustly detaining a ton of bar iron," a value of £30 was cited. In September of that year, however, bar iron and iron pots were being weighed out in Hutchinson's warehouse. Here the bar iron was valued at £21, the pots at thirty shillings per hundredweight. There is nothing with which to explain the discrepancy in the wrought-iron figures, and we lack the data with which to suggest, even crudely, what rates were normal in the

[10] Data on the new ownership group may be found in the following sources: Mass. Arch., 59/148, 160, 173; *Essex R. & F.*, II, 29; VIII, 39.

[11] *Essex R. & F.*, II, 97–98; Essex R. & F., IV, 57.

Savage-Hutchinson period. It is clear, however, that nothing in these stray data gleanings conflicts with the dismal estimate of the ironworks reached by the General Court before authorizing the establishment of ironmaking facilities elsewhere.[12]

Inevitably, one must ask why things went badly. The reasons are not hard to find. Some of them go back to conditions which have already been mentioned, the vagueness in title of ownership, the fractionalized assets, the difficulties in finding good managers, the plague of lawsuits. Others are but the current manifestations of problems which were constants in the ironworks' whole history, their seriousness doubtless accentuated by the prevailing confusions and uncertainties.

However well off certain of the new creditor-owners were, and several were prosperous indeed, by Massachusetts standards, common sense would have dictated caution in investing in ironworks which might at any minute be handed back to the Undertakers by victory in a lawsuit. Justice would probably have required that Savage and his partners be compensated for their interim outlays, but in a legal snarl of the magnitude generated by the ironworks troubles, who could be sure of justice, especially if the original Special Court verdict were to be thrown out altogether?

Similarly, though the Savage group held the best of the ironworks assets, there was in the rest much that had doubtless served to reduce operating costs in Leader's and Gifford's time. Timber costs had been high when only cutting and hauling were involved. If now the wood had to come from other men's land, the ironworks fuel bill would run higher still. Savage tried to cure the problem of fractionalization by purchase, in his own name, of many of the old Company's assets that had fallen into other hands, usually at bargain prices, and often with an escape clause in the deeds to guarantee his receipt of full purchase price plus the cost of improvements should the Undertakers recover their property within specified periods.[13] It is doubtful, however, if he ever managed to get all of the assets reassembled.

In all likelihood, too, the new management group had to wrestle with a labor shortage. It has been suggested that some of the workers stayed on. Others did not. Webb informed the Undertakers that

[12] *Essex R. & F.,* II, 29, 127, where "qt" is obviously a misreading for "Ot."
[13] *Suffolk Deeds,* II, 265–72; III, 3.

"three of or chiefest" workmen had left in December of 1653.[14] Subsequently, Becx offered to help Savage by recruiting new tradesmen, but all this was presumably conditional on acceptance of his generous proposal of joint operation of the ironworks. When the latter was refused, Becx could hardly have felt inclined to assist in this or in any other area, rather the contrary. Savage, we learn from Becx's letter, had had trouble finding workers on his own. In time, too, there was at least some exodus of workmen to the new plant at New Haven. How serious this problem was, is hard to estimate, although at one point it was alleged that it had been a man's refusal to work for Purchas which led that gentleman to have him brought into court for incivility to one woman and marriage with another one who was already wed![15] It is possible at least to suggest that here was the stuff of a vicious circle—irregular employment leading to departure to greener pastures, and this in turn leading to still greater difficulty in making iron, and so on and on.

Finally, production costs were high, perhaps excessive. Even if there had been no question as to ownership of the ironworks, no danger of their loss, it is highly probable that a point would have been reached when it made little sense to go on pouring money into Hammersmith, though it had been handed to its new owners on a silver platter. Again, clear-cut data are lacking. We know, however, that when iron was available, it was sold for more than £20 per ton. Unless the new proprietors were working on a basis of skim and skip, we may assume that their operating costs were so high that they could not abide by the court-imposed ceiling and make money. Whether this went back to the scattered assets, or to an unusually high wage bill, or to managerial inefficiency, we cannot tell. If we are to believe Hutchinson, however, the various creditor-owners who had made new advances of supplies to the works never made a profit on their investment.[16] Sooner or later, even the bravest souls among them, those most intrigued by the potential of mineral exploitation in New England, could only be expected to call it quits. Certainly, investment did in fact dry up and the ironworks go to wrack and ruin.

On the latter point we have expert and presumably neutral testi-

---

[14] *Essex R. & F.*, II, 83.
[15] Suffolk R. & F., 863.
[16] Mass. Arch., 59/159.

mony. Joseph Jenks, skilled craftsman, well familiar with the plant and with the iron trade, deposed in 1678 that in February, 1658, the furnace and forge at Hammersmith were worth but £800, and good for an annual rent of only £6. The rent of the ironworks farm he rated at £26! In his own telling words, "The said Iron works with furniture is wholly ruinated except the dam and water course." Two other men, Joseph Armitage and Thomas Newall, concurred in the Jenks situation estimate. Even Oliver Purchas, who could hardly be expected to be biased in the wrong direction, admitted that both forge and furnace were in a "run down condition."[17] Barring the turning up of what we should today call an "angel," this was the end of the road for Hammersmith.

The entrance on the New England ironmaking scene of William Paine, a prosperous Boston merchant, early in 1658, gave Hammersmith a new lease on life. What prompted his interest is not known. There is every reason to believe that he thought he knew a bargain when he saw one. If the works were in the sorry state just described, one might reasonably expect that some of the Savage group would have been so anxious to get out of the ironworks business as to sell their shares at marked discount from their paper values. If, on the other hand, an infusion of new capital and vigorous management could bring what had been a well-conceived plant out of the doldrums and into effective production of a still much-needed commodity, was it not worth the risk?

On February 3rd, 1658, Paine brought up the interest of Henry Webb, which the latter had acquired in satisfaction of claims worth more than £1300, for twenty-five tons of bar iron, payable in four installments, a quantity of iron to cover his interim expenditures, and half of anything over £500 which the "now Possessors" might get, either from the sale of the ironworks in England or from their redemption by Becx and Company. Since the bar iron was apparently rated at £15 per ton, this was by no means a bad deal for Paine, although that gentleman was also putting up a £1000-bond to guarantee fulfillment of the above conditions, as well as a pledge to assume all of Webb's risks in lawsuits filed by the Undertakers. In short order Paine purchased the rights of Sheafe, Stoddard, Tyng, and others, including Hutchinson's ironmonger uncle in London, and

[17] Suffolk R. & F., 1719; *Essex R. & F.,* VII, 27; VIII, 39.

emerged as owner of three quarters of the Lynn and Braintree plants.[18] The purchase price in these transactions is not known. Presumably, it was in no case excessive.

Oliver Purchas now became agent, apparently on orders from Paine, and, within ten days of the signing of the purchase of Webb's share of the ironworks, received Paine's directions to set on the woodcutters, hire a man to run the furnace, and arrange with a couple of landowners to take out bog ore.[19] With the directions went a note of judicious caution. Purchas was told to "Keep these things just to you, please, and only treat with them by the way," presumably lest word of the coming in of a new and rich partner raise wages and prices. Other data on the start of the new operations are lacking. We may assume, however, that with Paine's decision to "make the Works go lively," his assurance that Purchas would have "estate enough to do it," Hammersmith again became a beehive of activity.

Thomas Savage had not sold out his interest. The working arrangement between him and the newcomer, formal or otherwise, seems to have called for Savage's putting up one-fourth, and Paine, three-fourths of the ironworks' supply. With the plant in production, and apparently there was at least a good stock of sows to start with despite the "ruination," they should have divided the proceeds in the same ratio. During 1658, however, whether because Savage was once again way out front in the matter of returns from the ironworks, or because Paine was putting up much more capital, more than £1000 to Savage's approximately £159, the crafty Captain got less than a ton and a half of iron while the Boston merchant was taking more than seventy-seven tons. Perhaps Paine had Savage's ironworks operating-capital investment to catch up with. In any event, the balance was in time straightened out, and thereafter, if we may believe Purchas, both supply and division of produce proceeded in the ratio of three to one.[20]

The new arrangement seems to have worked well during William Paine's lifetime; at least, the courts were amazingly free of cases involving ironworks personnel and ironworks business transactions. So far as we can tell, and admittedly much of our evidence is circum-

[18] *Suffolk Deeds,* III, 137–39; Mass. Arch., 59/147, 165, 181.
[19] Mass. Arch., 59/179.
[20] *Ibid.,* 59/171, 174; *Essex R. & F.,* VIII, 39.

stantial, the simplification of ownership and the adequacy of supply by Paine and Savage did make a difference. The achievement is all the more remarkable if the plant had been as run down, the working staff as scattered, as some of the data for the pre-Paine period suggests.

There were still difficulties, of course. There was one legal action which seems to reflect local awareness that at long last people with money were again backing the ironworks. In June, 1660, Adam Hawkes, the ironworks farmer-neighbor, from whose bogs had come some of the ore first used at the plant, sued "Mr. William Paine and Company of undertakers for the Iron works of Lynn and Mr. Oliver Purchas, their agent," for trespass, more particularly for damming the Saugus River water so high as to flood his lands and well. This attack was easily turned by proving that the new management had bent over backwards to keep the water low, and demonstrating that Hawkes had had satisfaction from the old Company covering past and prospective damage by flooding. There was also a more or less routine case involving a failure to cut and coal a quantity of wood according to agreement.[21] In this period, however, there is no indication of a clash between the partners, or of clear-cut shortcomings in their agent, Purchas.

There is some evidence of shortage of skilled labor in the forge category. In February, 1659, Paine and Savage leased the forge at Braintree to James Leonard, finer. By the terms of the agreement the owners were to repair the plant to the tune of £40 or more and provide £100 worth of stock, all in return for Leonard's delivering them three tons of good merchantable bar iron a year. Assured one year's tenancy, and with that of two more dependent on the pleasure of all parties, the skilled craftsman pledged himself to make good the sum advanced for supplies, and to deliver up the forge in as good condition as it was when the repairs were effected. In all else, apparently, he was on his own, free to run the forge and dispose of its product as best he could. Since the bar iron was rated at £21 a ton, his straight rent came to only £63 a year, a by no means unreasonable sum.[22] Given the size of the partners' capital outlay on a plant whose fixed value was far from inconsiderable, one must wonder if anything but a scarcity of skilled forge hands kept them from working their

---

[21] *Essex R. & F.*, II, 210–11, 301.
[22] Mass. Arch., 59/181–82.

branch forge themselves. Conversely, of course, their willingness to enter into the agreement with Leonard indicates at least an anticipated need for additional forge capacity. The plant at Hammersmith must have been going well.

That such was the case is evidenced by surviving production figures for the Paine-Savage period. These are fairly extensive, although not free of chronological and other vaguenesses. There is no question as to the ironworks' turning out a good range of products. From the furnace were coming the sows which fed the forge, hollow ware, and solid castings, such as anvils, weights, and firebacks. From the forge came bar iron, still the main Hammersmith product. The works also ground out wrought ware, including such items as hinges, shovels, hoes, and spades. It is in the matter of total quantities in the various categories that the production picture is less than clear.

William Paine drew from Hammersmith, between the time he entered the partnership and his death in October, 1660, about 143 tons of bar iron, nearly 15 of hollow ware, some 6¼ tons of solid castings, and wrought ware to the value of £290.[23] Converting the first three items to monetary values by multiplying by going rates enables us to set £3750 as the approximate worth of Paine's total receipts.[24] If next we apply the 3:1 ratio mentioned earlier, and assume that Savage's portion, averaged out in time and by product type, came to one-fourth of the whole, that gentleman's share will come out at about £915. We also have figures on Paine's investment of operating capital. In February, 1661, William Howard, a witness to his will, and one of his heirs, deposed that William Paine had "disbursed for the Carrying on the said workes" from the time he bought up the shares of Webb and the others "the sume of 3432.05.00½ or thereabouts Errors excepted."[25] Assuming that this was three-quarters of production costs and that Savage had put up his proper quarter share, we may postulate that between February, 1658, and the fall of 1660, it had taken about £4575 to make about

---

[23] *Ibid.*, 59/152, 173. Lower figures for Paine's bar-iron receipts are cited in *Essex R. & F.*, VIII, 39. Unable to explain the discrepancy, I have chosen to use the larger totals given in a carefully drawn record of distribution of particular products in designated periods.

[24] The rates are as follows: for bar iron, £20; for hollow ware, £30; and for solid castings, £25.

[25] Ironworks Papers, 266.

£4665 worth of iron. Hammersmith, by this reckoning, was making money, but certainly not very much! One must wonder, indeed, at the modesty—or accuracy—of a statement that Oliver Purchas later attributed to old William, namely that he had "got enough by the Works in his days."[26]

When Paine died, his interest in the ironworks passed to his son, John, who continued to operate the plant in partnership with Savage, and with Purchas, to whose guidance John had been commended in his father's will, as agent. Under the younger Paine's tenure, Hammersmith went from a state of modest prosperity to utter chaos. Initially, at least, things went well. Soon, however, the ironworks was being plagued by its old nemesis, a shortage of operating capital, resulting from what Purchas called the "failing" of John Paine's estate.[27] What lay back of this failure is open to conjecture. John had inherited only a portion of what had been a very large estate. He was probably considerably less able than his father. He had something of a passion for matters mineralogical, a hobby which ran to money in seventeenth-century New England. Whatever the circumstances, his situation progressively worsened, and with tragic consequences for the ironworks.

The transition from vigorous but less than richly profitable activity to doldrums can be measured in surviving production data. John Paine, from his father's death until at least 1663, or possibly 1665, received only 113 T. 12 cw. 10 lbs. of bar iron, 7 T. 9 cw. 14 lbs. of hollow ware, 1 T. 13 cw. 15 lbs. of solid castings, and £25.11s. worth of wrought ware. This, expressed in money, amounts to about £2528. If we again fall back on the 3:1 ratio, we get a total Hammersmith product value, during this second Paine-Savage period, of approximately £3370. For this phase we have no cost data at all. To judge from actual developments at the ironworks, however, they must have exceeded the return.

Not only was John Paine's share of the proceeds smaller than his father's, but it shrank as time went on. So we may infer, at any rate, from surviving data on Savage's receipts, which carry a chronological breakdown from 1658 to, and including, 1663. The Captain drew almost 26½ tons of bar iron between February, 1658, and the end

[26] Mass. Arch., 59/173.
[27] *Essex R. & F.,* II, 271–74; VIII, 39, 199–200.

of January, 1661, some 15 tons between the latter date and March 24, 1662, not much more than 6 tons from then until the corresponding point in 1663, and about 10 tons in the remainder of that year. The "spurt" suggested in the last-mentioned figure is not explained. It may have been a matter of compensation for a marked and consistent slump in Savage's sharing in the hollow and cast ware. Although, between February of 1658 and March of 1662, Savage had taken about 7½ tons of hollow ware, in 1663 he got little more than two hundredweight. Almost 2½ tons of solid castings had reached him in the former period. During 1662, old style, he was down to 1 T. 13 cw., and in the rest of 1663 he got none at all. What held for Savage doubtless applied to Paine.

When we lump the two Paine periods together and come to grips with gross totals, it is clear that the ironworks production was considerably lower than what had been achieved during Gifford's term as agent. Over the years, the Paines, father and son, took 256 T. 1 cw. 3 qr. 16 lbs. of bar iron and 22 T. 6 cw. 2 qr. 6 lbs. of hollow ware, worth £5121.18s and £669.16s.6d. respectively. Corresponding data on their total receipts of cast and wrought ware are not available. Over the years, Thomas Savage received 58 T. 2 qr. 17 lbs. of bar iron, 11 T. 5 cw. 25 lbs. of hollow ware, and 4 T. 2 cw. 1 qr. of cast ware worth, as figured at going rates, about £1160, £338, and £103, respectively. It is impossible to determine the value of Savage's share of wrought ware. Applying the 3:1 ratio to the figures just cited gives us a total of £6405 as the worth of the Hammersmith proceeds, exclusive of wrought-iron tools and the like.

We have no data on costs over the whole period, no basis on which to try to work them up synthetically. Whatever the bill had come to, Hammersmith appears to have turned out something over £6500 worth of products over approximately six years. This is clearly well under Gifford's, as estimated in Chapter VIII. It is somewhat more impressive, however, when set against one estimate of fixed plant value, an entry in the inventory of William Paine's estate which carried three-fourths of the works at Lynn and Braintree at a value of only £800.[28]

The transition from one Paine-Savage phase to the other had not been easy. It was only by a narrow squeak that the partnership had

[28] Mass. Arch., 59/152, 183; *Essex R. & F.*, II, 273.

survived the death of William Paine. As early as March, 1661, at any rate, Savage and the younger Paine were filing cross suits charging illegal possession and seizure of ironworks property. These may have involved no more than an attempt to get legal clarification of a business arrangement which seems to have been pretty much *ad hoc* throughout its existence. Conceivably, some skulduggery had been attempted, or so one might infer from the fact of Savage's suing to recover what seems to have been all of the ironworks, despite the Paine title to three-quarters.[29] Since the issues were resolved out of court, the whole matter is open to conjecture. Of the fact that one fruit of Paine's death had been a partner's feud, however, there seems to be no question.

With some form of composition duly effected, Savage and William Paine set about running the works, and, as we have seen, operations on this joint basis continued through at least some portion of 1663. Then, however, Savage apparently withdrew his active support. We have no reason to believe that there was any failing of his estate. One must assume that a hardheaded man of affairs had concluded that if things had gone less than swimmingly when William Paine was backing the ironworks, they were likely to go from bad to worse in John Paine's hands. It is possible, of course, that the cleavage which had appeared in 1661 flared up again and that two men reached a point where they could not do business in tandem. Somewhat strangely, however, there seems to have been no formal dissolution of the partnership. Savage just stopped putting up money and, presumably, sharing in the little the ironworks continued to turn out.

Surviving documents make it plain that John Paine and Oliver Purchas then tried to make a go of Hammersmith on their own, and were in trouble as early as 1664. In January of that year, the financial pressures on Paine were so strong that he had to mortgage the bulk of the Paine share of Hammersmith to Samuel Appleton of Ipswich. Appleton had married Hannah Paine, and their children, by the terms of William Paine's will, were to receive £1500 on their coming of age or marrying. Now, apparently fearing lest his children be deprived of their legacies by John's slumping fortunes, Samuel Appleton took title to all that Paine owned at Hammersmith except the stock of supplies and iron on hand, and this to hold until such time

[29] *Essex R. & F.*, II, 270–71, 273, 275.

as John Paine or his heirs made good the £1500.[30] One obvious consequence of this was further fractionalization of ironworks ownership. Paine, Savage, and Appleton now held claims to portions of the Saugus River establishment.

What was to grow into a long-festering business problem also hit John Paine at about the same time. In December, 1664, the already embroiled young man agreed to pay Eliakim Hutchinson, as attorney to Richard Hutchinson of London, 10 T. 12 cw. of bar iron or its equivalent in sterling on or before the end of September, 1665, "in full payment for the interest of the saide Hutchinson in the Iron workes." Whether this was a piece of unfinished business inherited from his father or a development from a new and unrecorded transaction of his own is not clear. In any event, the terms just cited had not been met as late as the end of January, 1668, since at that point Paine and Purchas agreed to pay Hutchinson the iron in question "out of the first blast that (they) shall have from the Ironworks of Lyn except some inevitable accident doe happen that the saide Workes cannot produce such a quantity of iron."

Here, too, Paine and Purchas welshed, and not because of any "inevitable accident." In January, 1675, Hutchinson sued John Paine in Suffolk County Court. The issues pondered there and on appeal to the Court of Assistants were confused and confusing, but at no point did either side claim that iron had not been available. Quite to the contrary, there was testimony from Hutchinson that the plant was still standing, that Purchas had admitted that there had been a blast adequate to make ten tons of bar iron, that Paine had told him that enough had been made to pay him off, and that he wondered why he had not done so, and, finally, that "the said works have made many a tonn of iron since that time. . . ." Legal technicalities gave the verdict to Paine. The total proceedings seem to suggest that the young man was better at the law than at keeping promises and making iron.[31]

We have little data, in other words, with which to document the production claims which it had obviously been to Hutchinson's advantage to maximize. True, Oliver Purchas was licensed to sell "strong water" at the plant in 1666, and this normally meant that work was in process.[32] And some four years later, Purchas, bent on

[30] Mass. Arch., 59/186.
[31] Suffolk R. & F., 1362.
[32] *Essex R. & F.*, III, 351.

installing a new anvil block, acquired a tree that had long served as a boundary marker. A careful searching out of every wisp of evidence, however, has disclosed nothing with which to counter Purchas' picture of the latter days of Hammersmith. With the failing of his associate's estate, he claimed.

The Iron works went to ruin, which deponent, as agent, repaired, and became much indebted, which he always judged Payne's estate should have paid. The works depreciated daily and Mr. Jno. Payne was not able to supply or repair nor to pay him his salary. The works as they are now are of no great value.[33]

Here, obviously, is another epitaph for Hammersmith, and one whose general validity is attested by records of mounting debts and suggestions of declining production in old court files.

All surviving records are silent on the actual cessation of operations. We do not know when the furnace was blown out, when the last bar of iron came from the forge. Even conjectural dating is difficult. "Circa 1670," with the *circa* broadly interpreted, is the best we can come up with as the terminal point of major ironmaking activity. By June, 1676, we know, there were "neither utensils nor stock" at Hammersmith.[34] As to what had happened between the blast in 1668, mentioned by Hutchinson, and that date we can but guess. If our general diagnosis has been sound, however, one would expect that the interim had seen Oliver Purchas trying to do what he could with a deteriorating plant, and with next to no financial resources. He stayed on, that is certain, and did his best to make things move. If Hammersmith had to die, its slow grinding to a stop under the protecting hand of an agent who had been its faithful steward through thick and thin was by no means a bad ending.

And if, in 1676, Hammersmith had neither utensils nor stock, by 1678 there was nothing but real estate. In May of the latter year, the residents of Reading petitioned the General Court for a clearing away of the ironworks dam in the Saugus River. Pointing out that they had refrained from taking this step until the plant was "broaken up" and had "holly ceased to be occupied," they now sought to make

---

[33] *Ibid.*, VIII, 39, 254.
[34] Mass. Arch., 59/165.

it possible for the fish, that had been a "great refreshing" to them as food and as fertilizer, freely to come up into the rivers and ponds "wher they have ther Natural Breeding place."[35] The Court granted a hearing on the petition. No record of formal action has survived. Later developments, to which we shall shortly turn, make it plain that a most eloquent petition fell on deaf ears. There is no reason to believe, however, that this reflected a different ironworks picture from that which the Reading citizenry, long deprived of their "refreshment" by a bustling Hammersmith, presented to the magistrates.

The concluding chapter of the Hammersmith story is largely the record of still another morass of litigation. This time, however, it was only title to real estate over which men wrangled. John Paine died early in 1676, his affairs in chaos, and his estate in such straits that his creditors were forced to take 6s. 8d. on the pound in payment of their claims.[36] He had not cleared the mortgage to Samuel Appleton. In June, Appleton sued Purchas "for keeping him out of possession or refusing to yield him possession" of the Paine portion of the ironworks. The Court, despite the fact that Purchas seems to have been no true partner of Paine's and no part-owner, ordered him to hand over the property *or* £1500. The latter was out of the question, and Appleton took possession, by turf and twig, of housing, land, dams, sluices, rights, and privileges. Though much pressed by Appleton, Purchas insisted that there was then no iron on hand and no tools. Samuel Appleton the younger, who eventually occupied and improved the farm property, complemented this depressing version of the state to which the ironworks had fallen by deposing that "there never was works or stock since (he) came there."[37]

Even as the Appletons were taking over, however, Thomas Savage, now Major Savage, reappeared on the scene to take steps toward reinstituting his claim to the remainder of Hammersmith. Putting in one Samuel Stocker as his tenant, he ordered him to use and improve one *half* of the ironworks farm. The younger Appleton proceeded to make it plain that he would allow the tenant the use of one *quarter* and no more. There had been no formal division of the land. Savage's insistence that the Appletons undertake, after all these

[35] *Ibid.,* 59/133.
[36] *Essex R. & F.,* VI, 344–45; VII, 101, 354.
[37] *Ibid.,* VI, 166; Mass. Arch., 59/188, 189.

years, to protect him from recovery by the old Company had ruled this out. His offer to sell the Appletons his share for £200 had not been accepted, probably for the same reason.[38] Of necessity, there followed a joint tenancy that augured only more and better lawsuits.

They came, and, for complexity and bitterness, matched anything that we have seen earlier. In June, 1678, Savage successfully sued young Appleton at Essex Court for "entering upon and possessing and using his (Savage's) houses, land, and Iron Works at Lynn, cutting his grass, wood, timber and ruining the said Iron Works." The defendant appealed to the Court of Assistants, whose verdict also went to the Major. Appleton then proceeded to attaint the jury, made his claims stick, and won a reversal of the former verdict and costs of court. An appeal which Savage then carried to the General Court produced, in October, 1679, the Solomon-like judgment that Appleton was entitled to three-quarters of the property unless Savage could prove his claim to more, and that costs should be divided between the two contending parties.[39]

Even this was not satisfactory, however. An action of review brought by Savage in Essex County, in November of the following year, was nonsuited, and it was only after another and similar action had reached an impasse that peace came to reign between Savage and the Appletons. It hinged on a pact that had been reached voluntarily and which did justice, or rather more than justice, to Savage. The property was to be evenly divided, with Appleton making the division but with the Major taking the parcel of his choice and "paying £10 for the privilege." Either or both parties, moreover, might sue to recover lands which had been taken up in the interim by others. Ironically enough, at least one such action was entered by the late contending parties now turned allies![40] The pact was agreed upon in June, 1681. To such lengths had the "lawsuits and contentions" set off by the 1653 verdict against the Company of Undertakers carried!

An equally serious legal snarl grew out of some direct and extralegal action against the ironworks dam in May, 1682. In the dark

38 Mass. Arch., 59/146, 165–66, 175.

39 *Ibid.*, 59/162, 169–70; *Essex R. & F.*, VII, 26; *Mass. Records*, V, 246; *Records of the Court of Assistants*, I, 123–24, 133.

40 *Essex R. & F.*, VIII, 37, 123, 149, 195.

of night three attempts at breaching were put through, one abortive, one quite successful, a third which put the finishing touches on the accomplishments of the second. From our vantage point it is clear that at least two sets of people had appropriate motivation, the Reading people, whose petitioning had gotten them nowhere, and the Hawkes family, those ironworks neighbors long sensitive to the flooding of their low-lying lands. While Massachusetts courts at each level, save only the highest, sought to settle the question of responsibility, their efforts were fruitless. Unquestionably the dam was wrecked. We, like the members of the old juries, must conclude that the cutting was done by "a person or persons unknown." Since the Hawkeses were cleared of suspicion, though rather less than easily, we may infer that the perpetrators of the evil deed, nameless though they must be, hailed from Reading.

Limitations of space rule out even a summary of legal proceedings which were as rich in human interest as in complexities. It is the plant at Hammersmith which is our main concern. In the welter of surviving court records, only two directly significant items emerge. One is an account of the actual damage incurred. According to the Appletons, the break carried away much of the dam and undermined the rest. The rushing waters washed away soil and gravel, took out two bridges and a lot of fencing, and left much cornfield and pasture flooded. Even more serious, the newly deposited soil and waste so filled the river bed as to make the stream no longer navigable except to a point about a mile downstream from the old landing.[41] Another just falls short of telling us when productive activity at Hammersmith ceased. Two Scotsmen, formerly indentured servants, familiar with the works over many years, deposed that the dam had been maintained by the various owners "all the while that the workes went wich was above twenty years. . . ."[42] Would that they had indicated how much beyond twenty years! Neither they nor anyone else, however, said anything of damage to furnace and forges and their watercourses. Since, obviously, the Appletons were trying mightily to prove great loss, they would have mentioned such damage if there had been any at all. In the whole body of testimony, there is nothing which runs counter to the Reading petitioners' 1678 statement that the

[41] Essex R. & F., XXXIX, 57–1.
[42] *Ibid.*, XXXIX, 101–2.

ironworks had been wholly abandoned. Hammersmith, in other words, had become a farm.

The "declining years" of Braintree Works saw little more than the disposal of real estate whose title was enormously clouded. There was one significant difference between developments there and at Lynn. In no quarter is there evidence of ironworks production at either furnace or forge by the creditor-successors of the Company of Undertakers after the sequestration of its assets in 1653. The furnace was, presumably, long since well beyond repairing. There are no data which suggest that James Leonard actually went to work under the forge lease mentioned earlier; indeed, his Taunton ironworks connection should by all rights have been keeping him abundantly busy during the period in question. And if the forge was not worked during William Paine's lifetime, there is no reason to believe that it operated during that of his son. On the contrary, we know that in 1665 the latter suggested to Savage that they sell off the Braintree property, on the grounds that people were carrying off their wood and timber, and that "all they had there together lands and otherwise did bring them in no proffits as it is. . . ."[43]

With Savage's approval, John Paine proceeded to the disposal of certain of the Braintree holdings. In December, 1665, he sold Richard Thayer of Braintree the dam, houses, orchards, and other lands northwest of the Monatiquot.[44] In September, 1667, the same man acquired title to a dwelling house, charcoal storage house, orchards, and land on the north side of the river. Unfortunately, the hapless Paine neglected to insert in the deeds notice that he was the true owner of the property and would defend the grantee's title. This led to a pair of lawsuits in which Thayer's claims that he had been victimized were upheld. In one brought against Paine's estate, in fact, he was awarded damages of £102.8s.8d. and costs, four times what he was supposed to have paid for it in the first place.

The omission of the indicated items from the legal instruments was probably not accidental. Savage subsequently challenged Thayer's rights in the property, much as he tried to do with the Appletons' holdings of the land at Lynn. And, though it took at least two sessions of the Suffolk County Court and two or three of the

43 Suffolk R. & F., 1885.
44 *Idem.*

Court of Assistants during 1680–1681, he won his point. Although he had indicated that he would "stand to" anything that John Paine did "in point of sale or otherwaies," this was a case of selling land which he, Savage, owned. In the view of the Court of Assistants, at least, Savage held valid title under the old 1653 Special Court verdict; Paine was disposing of that which was not his; and Thayer was therefore no true owner.[45] At no point in the long-drawn-out proceedings did Savage claim that Paine had no right in the *sum* of the ironworks property. In one way or another, however, the property in controversy was held to be no part of that in which old William Paine had held a three-fourths interest. One must wonder why not, but vaguenesses in the bases of partnerships have doubtless given birth to stranger things. Few men could match Savage, though, in maximizing the return from the vaguenesses.

With John Paine dead, the Thayer property retrieved, and various personal holdings, those long ago bought up from smaller creditors of the Undertakers, somehow exempted from the partnership arrangement with the Paines, Thomas Savage emerged as the *de facto* owner of the bulk of the Braintree assets of the old Company. He and his heirs sold these off over the years, and at a very tidy profit. The Captain, for example, had bought Braintree Furnace and its two-hundred-acre plot for £35; in March, 1670, he sold furnace and land, with an additional fifteen acres, for £150. Similarly, his son, Ephraim, in December, 1682, disposed of twenty-four hundred acres, nearly two thousand of which Thomas had bought for less than £50, for £500. The Savage family seems to have fared well in its creditor's investment in New England ironworks—even making generous allowance for the nervous strain of facing, over many years, the threat of recovery by the Undertakers.[46]

The purchaser of the 2400-acre tract was John Hubbard, a merchant industrialist. On the banks of the Monatiquot, on land purchased partly from Savage, partly from a William Penn, who had bought up the holdings of two small ironworks creditors in 1657,[47] he operated a plant that consisted of "Iron Workes, Forges, Dam and Pond, fflume & Sawmill." This plant was apparently not the Under-

---

[45] *Suffolk Deeds*, XI, 308–309; Suffolk R. & F., 1885, 2053.
[46] *Suffolk Deeds*, XI, 308–309; IX, 307.
[47] *Ibid.*, III, 30–32.

takers' Braintree Forge given a new lease on life. The deeds of conveyance to Hubbard are detailed but include no mention of the old forge. Moreover, the list of units just cited is followed by the flat statement, "by me (Hubbard) Erected and made." This need not necessarily apply to all of the units, of course. We know, however, that even as Hubbard was completing his transaction with Ephraim Savage, he was buying of one Joseph Allen the right to build dams on the Monatiquot.[48] One must wonder why new dam building would have been called for, if Hubbard had merely been taking over an already established water-power-driven forge. Hubbard's activities thus seem to play no part in the story of Hammersmith and Braintree Works proper.[49] We shall return to them in the next chapter.

What, then, happened to Braintree Forge? Reluctantly, we must answer, "We do not know." This unit is by no means as easily to be brushed off as the furnace, since we know that it was in at least reasonably effective production in Gifford's time. The fate of the branch forge was closely tied to that of Hammersmith, to that of Hammersmith Furnace in particular. Unless crude-iron production at Lynn pushed the directly adjacent refining facilities, there would have been no need to carry sows and pigs to Braintree. Short of converting the Monatiquot River forge to a bloomery or direct-process operation, there was nothing to be done with it. Much was conspiring to make it an orphan. We know it needed repairs when Paine and Savage drew up their lease with James Leonard. If the terms of the lease were not met, and the repairs not effected, deterioration would have gone on unchecked. All forge maintenance was costly. Here there was no inducement whatever to make the necessary expenditures. So far as we can tell, the forge fell to ruin, apparently so completely that it failed to get even incidental mention in surviving wills and deeds.

In the 1640's men had gone to Braintree and to Lynn and attempted to bend Nature to their purposes. In some thirty years Nature had come close to reclaiming her own. Colonial proprietors had taken over and tried to make a go of the venture which a group of English capitalists had established. They, too, had failed. We have seen particular

[48] *Suffolk Deeds,* XIII, 361–62; XII, 341–42.

[49] Interpretation to the contrary, advanced by Braintree and Quincy local historians, bears no documentary evidence.

problems that the Undertakers and their agents, and their successors, had to face. In the recording of events and developments we have come upon data indicative of the "why" of what happened, as seen through the eyes of principals, minor actors, and bystanders. Almost without exception, however, we have failed, down to this point, to come to grips explicitly with the reasons for the failure of a most ambitious industrial enterprise. Since, obviously, it will not do to bury a corpse without inquiring into the causes of death, a far from easy task can be postponed no longer.

Looking at the first New England ironworks over the whole span of their operation, we may take certain fundamentals for granted. The young Bay Colony needed iron. The plants of the Company of Undertakers were able to supply iron of good quality and in moderate quantity. Under no one of the various proprietorships were the ironworks really profitable enterprises, and to all appearances they should have been. Back of the "why" are complex interacting factors more easily listed than weighted, particularly in the absence of quantitative data for this enterprise and the New England economy.

Fixed capital investment seems to have been adequate, and to have been put up in the first instance by people who knew their business. The first ownership group, the Undertakers, had a special problem in the adequate supervision and control of that investment. Despite much effort, it is doubtful if they managed to develop effective techniques of control of their overseas investment. Certainly, they did not develop them in time to stave off catastrophe. With little question, too, the Undertakers were rather naive in some of their basic assumptions; they failed to make sufficient allowance for differences in the economic facts of life between colony and mother country. Thinking that foodstuffs were plentiful and cheap, they acted as though wages ought to be low, ignoring the pressure on wages of free land and underpopulation. We have seen, too, that they wanted to get paid for their iron in money when money was in extremely short supply.

Their successors were, of course, Massachusetts residents who were directly familiar with all of the problems of doing business in Massachusetts. They could provide regular on-the-spot control of "their" investment. They faced, however, a difficulty which is, in part, the reverse of the Undertakers'. They were dependent on the Old Country for some supplies, refractory sandstones, for example, and for skilled labor. They had business associations, of course, but supplies doubtless cost them dearly, and labor recruitment was far from easy.

Clinching evidence is lacking, but from this distance it looks as though mismanagement had figured in the Undertakers' and probably in their successors' difficulties. Whether or not Gifford was an honest man, his bookkeeping was sloppy, his manner of living grandiose. Purchas, with the best talents and the best will, could not have run a taut ship in the midst of all the messy partnership arrangements under which he had to work. William Paine probably brought good business skills as well as money into the ironmaking operation. His son can only have been a problem to Purchas and to all who did business with Hammersmith.

In the total mass of data on the ironworks, it is a shortage of operating capital that stands out above all else. The Undertakers, and those who followed them, all decided in time that they would not or could not continue to advance money or supplies. This is but another way of saying that the ironworks could not even keep going out of profits. For this, two key factors seem to have been responsible. One was the high cost of production. Costs were high, and apparently through the whole period, for supplies, for transportation, and above all, as one would expect, for labor. In a normal situation high costs could have been absorbed in higher prices for the goods which were sold. This, however, was ruled out by the ceiling price imposed by the General Court. The second factor was the importation of iron from England. Between the one and the other the proprietors were literally squeezed. Ironically, the imported iron sometimes sold not only below what would have been a "fair" price on the bar stock ground out at Lynn and Braintree, but below the ceiling price as well.[50] Almost by definition, the ironworks could not have been

[50] This is an inference from admittedly weak evidence. Data on the prices paid by New England merchants for imported iron in the seventeenth century are, to the best of my knowledge, not available. The inventory of the cargo of a vessel wrecked in the spring of 1652 shows a quantity of iron rated at £18 per ton. The cargo was saved and landed. The value therefore presumably allowed for costs of transportation, and may have been a regular going price. (*Essex R. & F.,* I, 259–60.) Retail prices on small job lots and in the western part of Massachusetts were much higher—£28 in 1652–55. (Major John Pynchon's Account Book, MSS. [Connecticut Valley Historical Museum], *passim.)*

profitable just so long as "normal" trade between Old and New England, and the ceiling price, prevailed.

A significant question, that of scale of enterprise, is, in the present state of our data, unanswerable. It is possible that the Undertakers' enterprise was all out of proportion to the general economic level of the area the ironworks had been built to serve. We know, for example, that there were Massachusetts bloomeries which presumably turned a profit in the period under consideration. In theory, of course, the economies of the direct process of iron manufacture should have put the low-investment plants out of competition. In actuality, we cannot fairly compare the one and the other. The owners of the bloomeries, having got no lush subsidy from the authorities, were not bound by the ceiling price.[51] Particularly when the stock of imported iron was low, they could charge what the traffic would bear, and they could go in and out of production, as conditions warranted, with maximum ease. Hammersmith and Braintree Works, by seventeenth-century colonial standards grand establishments, took heavy maintenance expenditures even when they were idle. To get and keep them going took a lot of money, which only a steady return in profits on product sales could have made available on a sound business basis. It is any man's guess as to whether this could have been achieved if the ironworks had been operating in a freely competitive economy and with capable management.

Judged by the practical test of survival, both Hammersmith and Braintree Works must be pronounced failures. Even before they became extinct, however, they had begun to provide some of the impetus of other ironmaking ventures, and much of the trained skill without which these might never have been started. The making of iron in New England well into the first quarter of the eighteenth century owed much to men who had learned their trades at the pioneer plants. As these crumbled into ruins, they left a legacy, a human legacy, which in many ways atoned for the failures.

[51] It will be recalled that the failure of the later proprietors of Hammersmith to comply with the price ceiling had been cited by the General Court as grounds for authorizing the setting up of the ironworks at Concord.

13

Heirs of Hammersmith

THE QUITE WIDELY scattered ironworks to which many of the men of Hammersmith and Braintree Works and their descendants moved on were markedly different from the mother plants. All processed bog ores. Some received concrete support from local authorities. Three were, or were projected as, indirect-process operations. A far larger number, however, were bloomeries, most of them of quite modest scale. Almost without exception, too, they were the product of "American" investment and of local initiative, much of it aimed at speeding the growth of particular communities by encouraging local industry. So far as we can tell from highly limited data, they tended to serve narrower market areas. They were also, in certain instances, rather healthier business enterprises than the more or less artificial creations which had been the giant establishment of the Company of Undertakers. Some were highly successful. One was to prove almost amazingly long lived.

The earliest of these successor plants was the first to be set up in the neighborhood of Taunton, in the Old Colony. In the fall of 1652 the citizens of that town voted to invite Henry and James Leonard and Ralph Russell to join with a number of local residents in the erection of a "bloomary work on the Two Mile River."[1] All three were Hammersmith employees; how they came in contact with the Taun-

[1] Mass. Arch., 41/673.

ton folk is not clear. Further, while the arrangements were worked out with Henry Leonard, and while he and Russell were granted land "as encouragement," only James seems clearly to have moved to Taunton to help set up the forge and stay on as a permanent settler, sire of a mighty clan.

Leonard and, with little question, others from Hammersmith and Braintree, and perhaps others still whose points of origin are wholly obscure had the technical skills. Some twenty individuals, including two women, put up the necessary capital in £20-shares or fractions thereof. These were joined, in time, by an equal number of investors, mostly Bristol County residents, but including Governor Leverett and a few Boston merchants. Initial capitalization came to about £600.[2] The town came through with land grants and an authorization to take ore and timber from its common lands. It subsequently became a shareholder.

Eventually, too, the General Court of Plymouth Colony made concessions similar to, but less sweeping than, those granted the Undertakers by the Massachusetts magistrates. It confirmed the wood-cutting privilege. It granted exemption of workers from militia service. There seems to be only one instance of an attempt at control of activities at Raynham Forge, as the Taunton ironworks was called. This was a 1665 order "prohibiting bad iron to be made there." There is also at least presumptive evidence that the magistrates, again like their Massachusetts counterparts, had also to deal with the problem of workers' moral lapses.[3]

Thus was begun what Samuel Maverick described in 1660 as "a pretty small Iron-worke."[4] Surviving documents seem to render suspect the claims of Leonard descendants and local historians as to the centrality of James Leonard in the earliest phase of activity. The manufacture of iron began in 1656, at which time George Hall was clerk, and John Turner, probably the son of a finer at Hammersmith,

[2] J. W. D. Hall, "Ancient Iron Works in Taunton," *Old Colony Hist. Soc. Coll.* 3 (Taunton, 1855), 131–62; E. H. Bennett, "Historical Address," *Quarter Millennial Celebration of the Town of Taunton, Mass.* (Taunton, 1889), 42.

[3] *Records of the Colony of New Plymouth in New England* (12 vols., Boston, 1855–61), III, 176; IV, 98.

[4] *A Briefe Discription of New England and the Severall Townes Therein, Together with the Present Government Thereof,* from MSS. written in 1660 by Samuel Maverick and recently discovered in the British Museum by Henry F. Waters (Boston, 1885), 22.

was "working the forge." Leonard, at this stage, was probably but one of several workmen. In June, 1659, Turner quit, and in the spring of 1660 the shareholders leased the forge to Hall, Hezekiah Hoar, and Francis Smith for five years. These men then set up a subsidiary, or operating partnership in which they were joined by thirteen others, including James Leonard, holding a half share.[5] From this time, probably as Turner's successor, he served as "principal workman," subsequently conveying to his sons, and they to theirs, the rights to work the forge hearths in a fascinating example of the "inheritance" of a job.[6] For all we know to the contrary, the sub-partnership disappeared at the end of five years, and, one assumes, James Leonard's interest along with it. Certainly, Hall served a clerk for thirteen years, with a one-year break, and it was not until 1683 that a Leonard, Thomas, a son of James, assumed managerial responsibilities.[7]

Raynham Forge was a two-hearth bloomery. The two hearths, normally designated as "East" and "West," seem also to have been referred to as "finery" and "chafery." This suggests a division between roughing-out and finishing operations, with the two heating units working in sequence rather than in tandem. Both were used, in the early years at least, in conjunction with one big hammer powered with water from a pond created by damming the Two Mile River. Whether the hammer and other heavy iron equipment came from Hammersmith or from England is not clear. The forge building was presumably crude. There was also a charcoal-storage house and at least one house for the accommodation of the workers.

Surviving accounts give a good picture of operations, production, profit distribution, and the like. Carrying indications of noteworthy developments such as "The (Indian) War began," and "The Coal House burnt full of coals," and of expenditures for major repairs and replacements, they also cover, in considerable detail, ongoing "normal" business activities. Production ranged, in good years, from twenty to thirty tons annually, and brought in from £400 to £675. On these low totals, however, profits were often quite respectable. The partners had to pay a clerk's salary, the costs of digging and carting

[5] Hall, *op. cit.*, 135–37.

[6] Elisha Clark Leonard, "Reminiscences of the Ancient Iron Works, and Leonard Mansions, of Taunton," *Old Colony Hist. Soc. Coll.* 4 (Taunton, 1899), 51–65.

[7] Ironworks Ledger, Leonard Papers (Old Colony Historical Society).

ore, cutting timber, coaling, upkeep of watercourses, chimneys, and machinery. They paid the bloomers, or "makers" of iron, about £5.15s. on every ton. Nevertheless, to cite but a couple of samples, the holders of shares that had been valued originally at £20 got dividends of £3.6s. apiece in 1683, and £4.8s. in 1688. Back of a profit picture which is in sweeping contrast to that of Hammersmith must have lain high-quality ores and highly efficient operations. These factors doubtless also figured in the longevity of Raynham Forge, a unit which operated under one proprietorship or another, and in different types of iron processing, until 1876!

An eighteenth-century document refers to this small bloomery as "the Ancient Iron Works which begatt most of the Ironworks of this Province."[8] The description is sound enough. Taunton and vicinity became a major ironmaking center in the colonial period. In the proliferation of plants, the Leonards played a large and crucial role. With the money earned at his bloomer's trade, James Leonard bought up land in impressive quantity. He was also able, some time after 1666, to set up a one-hearth bloomery of his own on Mill River. This was Whittenton Forge, worked for about one hundred and fifty years, and in the possession of Leonards until 1807. A second and third hearth were apparently added in 1694 and 1737. From the latter year, higher-grade New Jersey ores were processed. James' sons, Thomas and James, erected, some time after 1695, in the North Purchase of Taunton, the present town of Norton, and with land grants and other "encouragements" from the proprietors, still another bloomery. This was Chartley Forge. Costing about £300, it was operated until close to the end of the eighteenth century.

Elkanah Leonard, Thomas' son, had an interest, although not a large one, in King's Furnace, erected in East Taunton in 1724. Zephaniah, grandson to the second James, after a family quarrel over ore rights, which was richly aired in courts and church, built a single-hearth bloomery on Mill River in 1739–40. About 1720, James Leonard, a son of the second James, bought land in what is now Easton, and built still another forge and put it in the charge of his son, Eliphalet. Originally named Brummagem, it was more commonly referred to as Eliphalet Leonard's forge. His son, of the same name, is supposed to have made steel at Easton prior to 1771, and this man's

8 Mass. Arch., 41/670.

275

son, Jonathan, set up steel furnaces in 1787 and 1808. Another son, a third Eliphalet, established about 1790 a forge, trip hammer, and nailer's shop, which went bankrupt in 1801, but became, along with other properties of the Easton Leonards, the basis of the subsequently richly prosperous Ames enterprises.

At this stage, of course, we are far from Hammersmith and Braintree, although still short of the complete story of the Leonard ironmaking ventures. This is not the place to handle what is in many respects the very epitome of the American success story. The Leonards rose in the world. They took on offices in government, militia, and church. In many, indeed in most cases, they acquired more and more property and became "first families" of the towns to which their work carried them. The constancy with which the men of the family stuck to the ironmaker's trade borders on the incredible. Perhaps this much, at least, may be in some measure chalked up as part of the Hammersmith legacy.

Interestingly enough, the Leonards seem to have forgotten their humble beginnings at earlier ironworks. Perez Fobes, a great grandson of the original James, obviously knew much that had been passed down as family tradition. He was, nevertheless, able to write of Raynham Forge that it was "the first America ever saw."[9] Clearly, Taunton and the Taunton area had wholly assimilated this branch of the Leonard family over the years in which they rose to become the principal ironmasters of Bristol County.

Since Raynham Forge had been set up in a separate jurisdiction, Plymouth Colony, no question of conflict with the Company of Undertakers' monopoly was involved. Before iron could be made under any other auspices in Massachusetts, however, that monopoly had to be canceled. It was, as we saw earlier, in response to petitions from Lancaster and Concord for authorization to set up ironworks. Without much doubt, these petitions were entered in behalf of local citizens planning enterprises similar to that which flowered as Raynham Forge.

Of the Lancaster effort we know little. Dreams of metal making had figured in its original settlement. Robert Child had been one of the original petitioners for the right to settle there. Joseph Jenks had been one of a "Company Intended to plant at Nashaway," as the

[9] *M. H. S. Coll.*, III, 170.

place was first called, who petitioned for the building of a bridge over Sudbury River so that they might go to work there in the summer of 1645.[10] Despite the claims of local historians, it is doubtful if any concrete results were forthcoming. Child's early embroilments with the magistrates probably ruled out any plans he may have had for ironworks at Lancaster. It is almost certain that Jenks never took up residence there. By November, 1647, several who had actually "planted" asked for permission to abandon the settlement, and while the petition was turned down, it clearly states that no real progress had been made.[11] By inference, the expectations for iron ores which had come to Winthrop's attention years before had not been realized.

In Concord, however, things worked out differently. Some months after the cancellation of the Undertakers' monopoly, a partnership for erecting of ironworks was formed. The identity of the first shareholders is not clear. Presumably they were mainly, if not exclusively, local residents. They were also apparently as eager for assistance from the authorities as the promoters of the earlier ironmaking enterprise had been. They asked for, but did not get, the right to dig ore in private property without the owners' consent. They got the right to do so in any of the common lands of Massachusetts. They seem also to have received from the town a grant of land on which the works was erected, and of plots for the use of workers.

At a point on the Assabet River, in the district subsequently known as Westvale, an extensive bloomery operation was begun about 1660. Judging from later deeds, the owning company was organized on a basis of thirty-two shares. By 1664 thirteen of these were held by John Paine. By 1671 five were owned by Oliver Purchas. In 1672 eight Concord residents and a man from Woburn held the remaining fourteen.[12] It is possible, of course, that the latter group had invited Purchas and Paine, or his father, to come in as "experts" earlier than the dates indicated, perhaps at the beginning. Conversely, two men who ought to have had their hands full at Hammersmith may have turned to Concord as a new field for investment in ironworks. We do not know. It is clear, however, that before the end of 1672 the Con-

[10] Mass. Arch., 121/5.

[11] *Mass. Records,* II, 212.

[12] *Suffolk Deeds,* IV, 148–49, 218, 237; Middlesex Deeds, V, 185–91; VI, 46–47; VIII, 212–18.

cord plant was wholly in the possession of two Boston merchants, Simon Lynde and Thomas Brattle.

There is no record of Purchas' having more than a financial interest in the Concord operation. Paine was more than a silent partner; at least, in 1662 he wrote a letter in which he tried to secure the release from jury duty of John Smedley, the ironworks clerk. Smedley was presumably in direct charge. No evidence as to the extent of his iron-trades experience or its source has been uncovered. The same holds for John Hayward, who was holding down the clerk's job by 1670. Neither, so far as we can tell, was ever associated with Hammersmith or Braintree Works.

It is not only *via* Paine and Purchas, however, that we count the ironworks at Concord related to the pioneer plants of the Undertakers. There was at least some continuity of workers. The order canceling the monopoly had stipulated that there be no raiding of the working force at Lynn and Braintree. Workmen could not be taken on at Concord without the consent of the owners of the older establishments. Despite the restriction, which incidentally may have been easily got around, given the Paine-Purchas interest at Concord, the younger Joseph Jenks, certainly, and the Scotsman, Thomas Tower, probably took up work on the banks of the Assabet. There may have been others from Hammersmith as well. Surviving data are so scanty, however, that we do not know, and perhaps never shall know, who did the real work at Concord.

While there seem to be no documentary data on production at Concord, the rather full spelling out of assets in the transfer of fractional shares gives us a picture of a quite impressive ironmaking plant. No reference to a blast furnace has been discovered. In all likelihood, therefore, the ironworks at Concord was a bloomery. The number of hearths is not specified but, presumably, there were at least two, and two hammers as well. The listing of dams, ponds, and watercourses in the plural suggests moderately extensive facilities. Land holdings were impressive. To a basic tract of 260 acres, 1000 were added by purchase, and another 260, by town allotment, all prior to the Lynde-Brattle ownership. Those gentlemen bought up at least 150 additional acres, so that what came to be called the Iron Works Farm included about 1700 acres spread over parts of the present towns of Concord, Acton, and Sudbury.[13]

Only stray data on the operating span of the Concord ironworks are available. Activity in the summer of 1660 is indicated by Jenks' taking on at that time iron from Hammersmith for use in the making of hearths.[14] In 1670 one Michael Wood got permission to sell "strong liquours to the labourers about the Iron workes . . . for their necessary releefe," and to no others.[15] A pair of indentures drawn up in 1682 as a basis for joint operations by Lynde and Brattle carried the statement that they "have equally improved together and as yet do."[16] Local historians have argued, apparently from the listing of "going gears" in a deed of sale, that the plant was in operation as late as the end of 1694.[17]

There is some reason to believe, however, that activity did not extend much beyond the death of Thomas Brattle in July, 1682. His interest in the Concord plant passed to his sons-in-law, in the right of their wives. In 1682 they sold out to Peter Bulkley of Concord and James Russell of Charlestown. Lynde died in 1687 and bequeathed his share to a daughter. Her husband, George Pordage, mortgaged the property it covered in 1689, and again in 1694, to Nathaniel Cary of Charlestown. Cary took title, presumably by foreclosure, and, in the spring of 1700, sold what had been the Lynde interest to James Russell. Bulkley's interest having been somehow eliminated in the interim, Russell thus emerged as sole owner—and promptly began selling off portions of the real estate.[18] Except for the "going gears," there is nothing which indicates the carrying on of the ironworks business by the Brattle and Lynde daughters and sons-in-law. The record of quite prompt sale and mortgaging seems to suggest the direct opposite of vigorously pressed business activity.

For the ironworks of Winthrop the Younger in Connecticut, the Undertakers' monopoly posed no problem. Here was a wholly colonial

[13] Middlesex Deeds, V, 185–91; VI, 408–10; VIII, 273–75.

[14] It was while at Hammersmith getting plates for the Concord forge that he spoke out against the King, saying he would cut off his head and make a football of it. (Mass. Arch., 106/28–34.)

[15] Middlesex R. & F., 1670–46–5.

[16] *Suffolk Deeds*, XII, 324–25.

[17] D. Hamilton Hurd, ed., *History of Middlesex County, Mass.* (3 vols., Philadelphia, 1890), II, 598.

[18] Details of these transactions can be found in the following sources: *Suffolk Deeds*, XIII, 96–97; Middlesex Deeds, IX, 70–75; X, 363–64; XI, 89–91; XII, 476–81, 634–35; XIII, 42–43.

venture, staffed in large part by old Hammersmith and Braintree workers, and intermediate in planning scale between the plants on the Saugus and Monatiquot Rivers and the comparatively simple bloomeries at Raynham, Concord, and other places. It was inevitable that Winthrop think in terms of the indirect process, that he start with a blast furnace. Here, however, he did not have the support of a rich English company, and grand plans were attempted with little more than local shoe-string capitalization. The result was a uniquely fascinating specimen of an infant industry born in trouble and carried through amidst varied complications legal, political, social, and technological.

In all likelihood, minerals figured large in Winthrop's thinking when first his main energies were focused on Connecticut about 1645. No sooner was he elected an assistant in that colony, than he received rights in any mines of lead, copper, or tin, and deposits of antimony, vitriol, black lead, alum, salt, or the like that he might find and develop in areas unsuitable for settlement.[19] Iron is conspicuously absent from the list, but by 1655, perhaps after a survey of the type he had earlier conducted in Massachusetts, it was on that metal that he was pinning his hopes.

In the spring of that year, Stephen Goodyear, a New Haven merchant, gave notice to the town that if any knew of iron deposits thereabouts, Winthrop was now on hand to test them. If the prognosis was favorable, an ironworks of "great advantage to the Towne" might be built.[20] In May, a letter to Winthrop from a trio of settlers in the Stratford area suggested that he send Goodman Post, until recently a key figure at Hammersmith, to check on some rock ores "that the designe for iron may bee promoted in these parts as soon as may bee."[21] There were doubtless rumors of other finds but no data on them, or on further developments at Stratford, have survived. By November, however, plans for ironworks at New Haven were firm.

In that month, Mr. Goodyear announced that he and Mr. Winthrop proposed to erect an ironworks "w(hi)ch is conceaued will be for a publique good," and asked the townsfolk to join in the undertaking. There was good reason for the men of New Haven to be interested.

[19] *Public Records of Connecticut* (15 vols., Hartford, 1850–90), I, 222.
[20] F. B. Dexter, ed., *New Haven Town Records, 1649–* (vols., 1917–), I, 235.
[21] Winthrop Papers (Connecticut State Library), items dated 22 May 1655.

A plant which would bring not only iron but people and trade as well could do much for a settlement whose rather ambitious plans had not worked out too well.[22] For all this, however, there was no little reluctance to support the project, much of it going back to doubts about the character and assimilability of the workers. The problems to which they had given rise in Massachusetts were apparently all too well known! Money, of course, was tight, but those who concluded that the assets of a new industry would outweigh its liabilities quickly offered a substitute, their own labor. Joint efforts of fellow citizens in barn raisings and the like became commonplaces of American frontier rural life. Here they were to be lavished on an infant industry. More, much more, was to be required, but the first town-meeting consideration of the Winthrop-Goodyear project produced pledges of about one hundred and forty man-days' work.

Within a fortnight some had made good on their pledges but others had not. Another town meeting was therefore called to consider how to persuade the recalcitrants to perform, and also, and more important, to find out who would advance money. There was no rush to get aboard the new ironworks bandwagon, and it was left that individuals who might be interested were to meet with the governor, Theophilus Eaton, in private session in the afternoon. The town meeting proper managed, however, to offer some public encouragement of the new venture. It voted permission for the ironworks to go on, and granted rights in wood, water, ore, and shells in the town's common lands on the East River, provided that no man's property be entered upon. Also, others were permitted to cut wood there in an orderly way. And it was voted that Branford make similar concessions.

Branford was involved because the proposed site was roughly athwart the still less than clearly drawn line that separated the two towns. It duly voted its equivalent of the grants from New Haven, apparently to Winthrop directly. Eventually, it was arranged that five-eighths of the wood and "other things of like nature" needed by the ironworks would come from New Haven, the rest from Branford.[23] The total offerings from both towns were far from generous, all things considered. In addition, however, the promoters of the project were to benefit from a rather nice concession from New Haven

---

[22] *New Haven Town Records*, I, 260–61, 330–31; Winthrop Papers, 15.113.
[23] *New Haven Town Records*, I, 349.

Colony. It had voted some months earlier, and doubtless under the influence of John Winthrop's eloquence, to grant exemption from both personal and property taxes on ironworks established anywhere in the jurisdiction.[24]

With little question this was by plan to be Winthrop's ironworks. Not only was Goodyear initially acting as his agent and spending his money, but, according to his own statement, he had first tried to get local residents to *lend* money against the day when they would be repaid in iron. Finding that none would lend but that some would invest, he settled, in November, 1655, for an arrangement which had ownership split in quarters, one to residents of New Haven, one to residents of Branford, and one each to Winthrop and Goodyear.[25] With little question, too, it had been a desire to share in the *control* of the ironworks that had prompted some of the members of the most Puritan of the Puritan settlements to insist on ownership rights, and this despite the universally high esteem in which Winthrop was held.

Trouble would probably have arisen had there been one town making the appropriate grants, and one town "bloc" of presumably small shareholders. The fact that there were two of each more than doubled the potential. And Winthrop, one might almost say, characteristically, had rushed into construction under a thoroughly murky legal title. By December, 1655, the dam, built at least in part by the labor contributions of the townsfolk, was fourteen feet high, major expenditures were being made, and it was not established whether the plant site belonged to New Haven, to Branford, to the investors in the two towns, or to Winthrop!

Worse still, besides insisting on half ownership in the ironworks, the residents of the two towns seem to have tried to get in on construction activities, all on their own and without so much as a by-your-leave from Winthrop or his agent, now turned partner, Goodyear. Details are lacking. We know, however, that they took direct action of some sort as a result of what John Davenport described as "an hurry of temptation."[26] Whether prompted by a desire to stake a claim to what promised to be a good investment, or by suspicions as

[24] *Records of the Colony or Jurisdiction of New Haven, from May, 1653, to the Union* (Hartford, 1858), 149.

[25] Winthrop Papers, 13.14, 114.

[26] Winthrop Papers (American Antiquarian Society), 121.

to the arrangement in force between Winthrop and Goodyear, "too many hands" came on the scene, and greatly to the embarrassment of Goodyear, who had twenty men at work, including some Frenchmen he had recruited at Manhattan, and four hundred cords of wood and a partially completed dam to show for his pains. Clarification of the whole basis of ownership and operation of the ironworks was obviously in order, if iron were ever to be made at New Haven at all.

The product of negotiations toward this end was a pair of documents signed in February, 1656. In the first, Stephen Goodyear, apparently on his own, and Jasper Crane and John Cooper, presumably as agents of the investors of Branford and New Haven, respectively, agreed to finish the furnace at their own cost. Winthrop was to be allowed a quarter interest in return for his "discovery," his securing of privileges and immunities, and his promotional costs. He was not to share in the first yield, however, and might thereafter only if he contributed his fourth of the costs of operation. The other instrument called for the erection of a forge on the same ownership basis. With it, however, went an interesting proviso which enabled any of the four, once the furnace had been worked for two years, to build forges of their own elsewhere than "upon the ground appropriated to the furnace."[27]

In May, 1656, additional official grants were made to, or in behalf of, the ironworks. In one, the New Haven General Court allowed twelve acres of ground to a collier for the works, and indicated that the newly settled proprietorship had not dispelled all need for caution as it did so. The land was to belong to the collier "if the Iron worke goe on" and if he stayed three years. It was also made plain that he was to abide by the town's orders.[28] The second, more than a little unusual, required creditors who attached any man's share in the ironworks to bear his responsibilities toward it until such time as their claims could be made good out of its proceeds.[29] Not even the administration of justice in civil proceedings was to stand in the way of progress toward the much desired end of ironworks for New Haven.

Furnace construction, begun at a point near the outlet of Lake Saltonstall, on what is now the main road between New Haven and

[27] Winthrop Papers, 13.14, 115, 117.
[28] New Haven Town Records, I, 279.
[29] Records of the Colony or Jurisdiction of New Haven, 173.

Branford, and originally estimated to cost about £2000, was plagued with difficulties. Winthrop had had to ask Richard Leader for direction as to how to go about building a furnace hearth, and had been referred to William Osborne, an experienced man, who in time turned up at New Haven, doubtless to lend a hand at the ironworks. There had been a problem of finding a suitable clay, a problem not solved before August, 1657. More important still had been the trying business of securing suitable hearthstones, a matter in which a number of people were concerned, and for which there was no satisfactory conclusion prior to November, 1658, and probably not for some time thereafter. The ore was not wholly pleasing. Roger Tyler, the Hammersmith founder and no novice in these matters, had found it "very meane."[30] All of these problems were in time overcome. It is doubtful in the extreme, however, if New Haven Furnace was in effective production before 1663.

Long before that time additional changes in the proprietorship had occurred. Goodyear and Winthrop lacked the capital even to keep the workmen adequately supplied. Difficulties of the sort just summarized doubtless put a damper on the willingness of local residents to put up additional funds. Further, Winthrop's election to the governorship of Connecticut in May, 1657, all but guaranteed his absence from New Haven, and the loss of his influence and persuasion that must have done much to ease relations among the ironworks partners. With all this in the immediate background, Stephen Goodyear tried to get out. In August, 1657, he asked the town of New Haven for permission to sell his share of the business. Told that he might not, except to parties approved by the town, he took off for England the following month, presumably to raise more capital. In this he was apparently wholly unsuccessful.

Winthrop fared better. In the fall of 1657 he leased his share of the ironworks to William Paine and Thomas Clarke, another Boston merchant, for seven years. The fact that he was by then resident in a different jurisdiction made it possible for him to do so whether the men of New Haven liked it or not. Clearly they did not. They saw a threat to local trade in the coming on the scene of a pair of rich Massachusetts merchants. They also feared that "A disorderly Company of workmen may be brought which may be to the dishonour of

[30] Winthrop Papers, 13.116.

God, And will bring mischeif upon the place; the worke alsoe."[31]
They had no choice but to concur in the new arrangements. In doing
so, they made it abundantly clear, among other things, that the grants
of public encouragement applied to the furnace and forge only within
limits approved by the Court, that no part of the property could be
sold or leased without express permission, and that all employees
were to be subject to all the regulations of the town.

If caution continued to reign among the ultra-Puritans of New
Haven, the two newcomers were tending strongly in the opposite
direction. From a letter which Paine and Clarke wrote Winthrop asking him, as true owner, to support the efforts which Jasper Crane, at
their suggestion, was making to get additional grants and clarification
of old ones from Branford and New Haven, we can tell at least what
they wanted. They had had the temerity to ask for such things as
title to lands flooded by the building of dams, and to upland and
meadow near the plant for the use of workmen and animals; the right
to take ore, and large timber not available on their own land, wherever
they might be found; and highways for hauling coal and ore to the
furnace, and coal and iron to the forge. They also sought, properly
enough, to have the ironworks' timber-land grant recorded and to
work out some form of operating agreement which would be equit-
able for all parties concerned, but which would also guarantee that
"the neglecte of some may not preieduce the Rest."[32] Crane's per-
suasiveness seems to have worked at Branford; at least, that town
made additional grants to the ironworks in the spring of 1658. It is
doubtful if New Haven saw fit to extend, now, and to outsiders, more
than it had been willing to offer Winthrop when it was bending every
effort to persuade him to settle there.

The new partners did more than ask for privileges. They put up
money, and apparently in sufficient quantity to develop the forge and
get the furnace into real production. Proper dating of the wrought-
iron unit is impossible. Doubtless projected from the beginning, in all
likelihood it had not been erected by the summer of 1658. Whenever
built, it could hardly have been worked until the furnace was operating successfully. The layout of the forge plant is even more obscure,
although the resources and interest in matters metallurgical of Paine

[31] *Ibid.*, 15.113.
[32] *Ibid.*, 3.111.

and Clarke ought to have guaranteed that it be well built and equipped. Given the background of financial and technical difficulties, it is quite unlikely that the New Haven ironworks could have become a going concern without their help.

The basis on which the business operated is less than clear. Stephen Goodyear died in the autumn of 1658, still possessed of his share in the ironworks. It is possible that Paine and Clarke bought or leased it from his estate. Whether or not these men also took on the interests of the local residents is not known. Conceivably, inability to match the outlay of capital by the wealthy Bostonians caused them to withdraw or to settle down as wholly silent and powerless partners. The Paine-Clarke lease from Winthrop presumably ran to 1663. In 1660, however, William Paine died, and the interest in the New Haven works, which he had held with Clarke, passed to his son, as did one in Winthrop's graphite operation at Tantiusques.

As to how well the ironworks fared while the elder Paine lived, we are not informed. With little question, however, his death was a catastrophe for the Connecticut, as well as for the Massachusetts, ironworks. As early as 1661, John Paine was writing Winthrop, now asking assistance in finding much-needed forge hands, now begging for more time, presumably to make good on payments due him, and eventually suggesting that neither he nor Clarke would object to Winthrop's reclaiming his old share. They appear to have carried on, perhaps under a new arrangement with the Connecticut governor, but, by 1665, Paine appears to have disappeared from the scene. Clarke, at any rate, was by then "Master" of the ironworks and whole owner of at least the ironworks "Farm." From that time to the point when he sold the farm in 1680, and by then the plant was out of operation, it was he who ran, if he did not own, the whole enterprise.

All through these years of failures and successes, workmen had been on hand. In the original nucleus were Goodman Post and a number of Scots, of whom the men of New Haven took a dim view indeed. Eventually, a quite large number of men who had earlier been employed at Hammersmith and Braintree, and various members of their families, went to New Haven. Among them were John Hardman, Nicholas Pinnion, Ralph Russell, Roger Tyler, and John Vinton. There was also William Osborne, clerk at the New Haven works until his death in 1661. Some of the workers had no connection with the

Undertakers' old plants, Goodyear's Manhattan Frenchmen, for example. More "alumni" of the first Massachusetts ironworks gathered at New Haven, however, than at any of the other ironmaking plants of that particular period.

On the workers' normal and routine activities we have only fragmentary data, inventories of estates, records of civil suits, and the like. The record of their sins and shortcomings is as extended as it is pungent. Now there is a case of an ironworks clerk accused of trying to violate the chastity of Pinnion's daughters, now one in which a Pinnion son is found guilty of contemptuous speeches against the authorities, now one involving breaking and entering Clarke's house and making off with merchandise *and* the book in which the workers' debts were carried, and, early and often, incidents of drunkenness, Sabbath breaking, assault and battery, swearing, slander, and defamation. Even amidst Puritans who considered their Massachusetts coreligionists lax, the men who made iron, and their womenfolk, were clearly refractory in the extreme.

The authorities tried their best. In August, 1665, the town of New Haven pondered the "disorderly persons that were at iron-workes," voted that no person who lacked a certificate of good character could be employed, and set quite heavy penalties for noncompliance with the order.[33] From 1666 to 1680 a special ironworks constable was appointed annually. Neither the required endorsement as to "civil" life and "blameless conversation," nor the presence of what one might be tempted to call a special supervisor of ironworkers' morals effected great change in people described by one of the key figures in the community as "very chargeable & froward."[34]

Hopefully the workers turned out enough iron to compensate for the stresses and strains which they thrust on the community. No production data have come down to us. It is difficult, indeed, to tell how regularly and how long the plant was in operation. Permission to dispense liquor was granted in at least 1661, 1662, and 1666. Activity was doubtless carried on in other years. In 1669, for example, the ironworks got a seven-year extension on its tax exemption. This would hardly have been granted without expectation of continued production and benefit to the region. There is no record of a second extension, probably because men were doubtful of the plant's future.

[33] *New Haven Town Records,* II, 146.
[34] *4 M.H.S. Coll.,* VII, 524.

On two occasions the authorities formally allowed the proprietors more time in which to meet their obligations to their creditors. Of the first "indulgence" we know only that it did not produce the "desired effect on the accounts of the iron Works." The second came in 1678, and in response to a petition setting forth Clarke's losses "by the late tremendous Indian warr" and others in prospect for the ironworks "by debts and other wayes."[35] Clearly, there were financial difficulties. It is also likely that Clarke was plagued by the deaths of key workmen. By 1679, Tyler, Russell, and Pinnion were all dead. Their replacement could not have been easy, even if adequate capital was available.

Probably because of factors such as these, Clarke abandoned the making of iron about 1680. He seems to have excepted the plant proper from the ironworks property which he conveyed to John Potter in August of that year. By the following July, however, the furnace dam and watercourse were granted to Samuel Heminway to set up a gristmill. Three years later the same man got the right to erect a fulling mill on the site of the ironworks forge. Neither of these grants would have been voted had there been any hope of resuming production in iron. Potter, it is true, intended to build a bloomery and make iron by the simpler indirect process. With Thomas Pinnion, he petitioned for the right to do so in 1692, and was given land, ore rights, and assurance of water from the furnace pond, despite Heminway's proposed gristmill. There seems to be no evidence that his plans were carried through. Similarly, the gristmill appears not to have been erected, and it was only in 1709 that the fulling mill, set up by Heminway's sons and John March, arose where once Clarke's forge had stood.[36]

The ironworkers, and their children, settled down mainly as farmers, joining the church, sharing in the distribution of lands, and coming, in general, strongly to resemble the regular inhabitants for whom they had once been trial and tribulation. At New Haven, as at Lynn and Braintree, agriculture and the dominant mores triumphed over industry and the tough raw material of early industrial workers. A crumbling furnace, a forge converted into a fulling mill, and whatever lingered in men's memories—to this had New Haven's ironworks

[35] *Public Records of Connecticut*, III, 4–5.
[36] Stephen Dodd, comp., *The East–Haven Register* (New Haven, 1824), 27–31.

come in about twenty-five years of vigorous public and private joint enterprise.

Another New England ironworks with Hammersmith roots seems to have gone little beyond the project stage, despite the expenditure, over several years, of much effort and capital. This was the plant promoted and partially carried through by John Gifford in that section of Lynn which is now North Saugus. There, in 1662, the ex-agent of the old Company of Undertakers bought about two hundred and sixty acres of wilderness land for less than £200, cleared and broke up a portion of it, set out an orchard, and built a house.[37] Despite these farm activities, and doubtless because his interest in mineral prospects had been in no way dampened, his pride much injured by his grim experiences as agent and thereafter, it is more than likely that Gifford intended from the start to set up a rival to Hammersmith. By the beginning of 1663, at any rate, a company had been formed and Gifford was back in the iron business.

Little is known of the company, but in such data as have survived we can see Gifford operating with all of his old audacity and with considerable success. The site was his. He could claim years of experience. True, he lacked capital, but that is what partners were for. In October, 1662, came the mortgage of the farm to Thomas Breedon, a Boston merchant, for £200 payable in September, 1664. The following January, he contracted with Breedon and John Paine to set up works for producing iron ware *and* other metals, the latter apparently including silver. It may have been rumors of this precious metal which led to arrangements which would impress a modern financial wizard. Gifford was to put up half of an initial stock of £600., hold a half interest in the works, and draw an annual salary of £150. His "investment" he borrowed from "Capt. Tho. Breedon & company," presumably the group in process of formation. Until it was repaid, from his share of the profits, his interest in a plant clearly specified as yet to be erected stood mortgaged, and the Company relieved, quite graciously, one might say, of the necessity of paying his salary.[38]

What must have been high hopes indeed were dashed, and comparatively quickly. Gifford promptly went ahead with construction activities, probably incurring more personal indebtedness to Breedon

---

[37] *Essex R. & F.*, III, 306; Mass. Arch., 39/673.
[38] Essex Deeds, II, 113–15; *Essex R. & F.*, VII, 176–77.

as he did so. By the winter of 1665–1666 that gentleman sued him, presumably for debt, and was awarded the quite substantial verdict of £1050.[39] Gifford tried to put his land into the name of his wife and daughter but was unsuccessful. It went to Breedon. So, one must infer, did whatever interest Gifford held in the "new Iron works at Lin." From a much disputed appraisal of the latter, rendered in February, 1666, we can get some notion of how far work had been carried. A dam, a furnace, and a dwelling house were finished and, only quite recently, paid for. These units were worth, according to Breedon's appraiser, £110, £350, and £140, and, according to Gifford's £250, £800, and £200, respectively.[40] There is no reason to, believe that they formed a really going concern; indeed, it was doubtless the complete absence of profit that had kept Gifford from making good on his commitments.

The ex-partner probably stayed on as works manager, and may have managed somewhere along the line to make at least some iron. No improvement in the profit picture can have developed, however, since, in December, 1669, Breedon, getting ready to return to England, handed over his interest in the plant to his late adversary in bitterly contested litigation, and at the bargain price of £400, payable in equal installments from the first and second blasts. In April, 1670, however, Gifford did better still in buying out Breedon altogether for only £200. With Paine's interest somehow liquidated in the meanwhile, the promoter of the ironworks venture emerged as its sole owner.[41]

In 1672, doubtless after unsuccessful attempts to work the plant on his own, and after fruitless efforts to interest local investors, Gifford sailed to England on a capital-raising venture. By his own statement he was approached by the representative of a group of capitalists interested in hearing his story. One suspects that this was no matter of coincidence. It is easy, indeed, to summon up a picture of Gifford talking freely in a London hostelry, now of lands of his known as the silver mines, now of deposits of gold, lead, tin, iron, and copper, now of a patent to one-fifth of all the metals to be found between Kennebec River and New York that he was about to receive.

[39] *Essex R. & F.*, VII, 175; Mass. Arch., 39/693.
[40] *Essex R. & F.*, III, 305–307.
[41] *Ibid.*, VII, 175.

It was half of all this, in any event, which Gifford sold to eight men, including a knight, Sir Richard Combe, and an author and translator, who was also a member of the Worshipful Company of Skinners, John Bulteel.[42] Wonder of wonders, he even persuaded his new partners to agree that each shareholder should pay "one Ginny peece of gold to be layd out in a peece of plate for the Said John Giffords wife for her Consent according to the Custome of New England."[43]

The new investors' enthusiasm did not stand in the way of their working out quite strict business arrangements. Gifford was to receive £1000 for the half interest he was selling, but he actually got only £80 while in London. The remainder, split into £420, which was to go to Gifford outright, and £500, which was earmarked as his share of the working capital, was to be payable only after he had clarified his title, brought the works to good repair, and handed over possession to Ezekiel Fogg, the man who had first approached him and who was now to return to New England and serve, jointly with Gifford, as agent pro tem of the English shareholders. Fogg and Gifford were to manage the company's affairs, with the former holding the purse strings and with both operating under written instructions, until such time as one John Wright could sail and take over as "superintendent" of all operations. Gifford's role was quite clearly to be secondary, and circumscribed. All this was worked out in August, 1673, two months after the drawing up of the original instrument of sale.[44]

With his newly found resources, Gifford seems next to have turned to a problem that had probably long been plaguing him, that of recruiting workmen. Record of only one contract has survived. In August, 1673, Wright, Fogg, and Gifford took on, as "master workmen for pot making," Henry Dispaw and a son of the same name. Bound for six years, and with their services at the employers' option thereafter, they were each to get £35 a year, housing, a garden plot, and passage, including that of the younger Dispaw's wife and child.[45] If there were others, one must hope that they did not give rise to conflicts like those generated by the arrangements with the Dispaws, to which we shall shortly return.

[42] *Ibid.,* VII, 175–77.
[43] *Ibid.,* VII, 117.
[44] *Ibid.,* VII, 173–74; Mass. Arch., 39/693.
[45] *Essex R. & F.,* VI, 82.

Letters went home to Massachusetts, some of them sufficiently glowing to persuade the Giffords that the man of their daughter's choice was no longer good enough for her to marry, others indicating further delay in the promoter's homecoming. Back on the scene, Gifford fell to with vigor and took on a stock of coal and ore, erected a building in which the potters could live and work, repaired bridges and dam, and reconstructed what appears to have been a seriously defective furnace. All this took the work of twenty or thirty men over six to eight weeks. By then four hundred loads of coal were on hand and the founder, unfortunately not designated by name, had been sent for. The clouds in Gifford's title that went back to his attempt to sign things over to his wife and daughter had been cleared to the satisfaction of the highest officials in the colony. All looked promising in the extreme.

Unfortunately a major clash between Gifford and Fogg developed. Much of the work just mentioned had been undertaken on credit. Fogg, for reasons which are not clear, refused to pay the bills. The storm broke in January, 1675, when one John Floyd, as attorney to the Dispaws, sued Wright and Company for their relief, and was awarded a judgment of £500. The judgment was executed against Fogg, who went to jail. Since the Dispaws were still in desperate straits the General Court ordered a new execution, this time against Gifford.[46] He, too, went to jail. Both confinements were long drawn out. Both came, it was alleged, just as the ironworks was about to go into production. At about the same time, moreover, an Indian uprising threw the colony into confusion, and much frightened workmen could not be persuaded to stay at the plant. From the piling up of these catastrophes the second ironmaking venture in Lynn never recovered.

There followed, of course, a maze of litigation comparable to that which kept Hammersmith and Braintree Works in various courts long after their abandonment. With Gifford as central figure it could not have been otherwise! Here, too, there were indignant English capitalists to deal with. In June, 1676, Wright wrote Fogg and Gifford asking them to "lay aside all animosities and consider if you can finde any expedient to advance our Common Interest." By October, doubtless informed that reconciliation was impossible, he gave power

[46] *Ibid.*, VI, 81–82; *Mass. Records*, V. 35; Mass. Arch., 39/690; 59/129; 100/190–91.

of attorney to Thomas Walter and Richard Middlecott to bring both Fogg and Gifford to account. While these attorneys were at one point authorized to treat with Gifford and try to arrange for a blast at the ironworks, with the attorneys putting up the wherewithal if the prognosis seemed favorable, cooperation with the attorneys was as futile as it had been with Fogg. In April, 1679, and in March, 1680, Walter and Middlecott sued Gifford for breach of covenant. Eventually they won their case. Gifford went to jail and remained there, except for a brief interval of freedom and numerous court appearances, until November, 1684, a matter of more than four years and seven months.[47]

Imprisonment put no damper on Gifford's ardor for legal controversy. There were suits and countersuits. There was question of everything from Walter's right, as a "stranger," to sue in the regular courts to the validity of his and Middlecott's power of attorney. There were petitions to the General Court in which Gifford was angry, indignant, humble, and pitiful by turns. The various legal papers of one kind or another reached mountainous proportions. For all these strenuous efforts, however, it was time which gave Gifford the ultimate victory. Walter, his principal adversary, eventually relinquished his attorneyship—and the battle. This development, and charity, led the General Court, at long, long last, to order Gifford's release.

Even then, Gifford was not ready to call it quits. As late as September, 1685, he was suing Walter and Middlecott in an action of review of the case he had lost to them in March, 1680. In January, 1688, he apparently won out in a case against them before the Superior Court of Appeals. Ironically, though quite appropriately, his vengeance reached out even from his grave. In the first year of the new century, Middlecott was ordered, on the petition of Mrs. Gifford, widow and executrix, to show cause why he should not pay the costs and damages assessed against him in the action just mentioned.[48]

What was happening at the ironworks through all this legal furor is almost wholly obscure. The claim that it was on the verge of production when the Dispaw case broke was often repeated. There is

[47] *Essex R. & F.,* VII, 171, 172, 180–81, 339, 341–42; Mass. Arch., 39/764; 40/190; *Mass. Records,* V, 225–26, 331, 455–56.
[48] Essex R. & F., IX-2 (July, 1684–March, 1686), 2; Suffolk R. & F., 4874.

only a wisp of evidence to suggest activity during the Fogg-Gifford joint managership.[49] A quite detailed appraisal made in the spring of 1679 listed land, dwelling house, the potters' shed, and the like. The furnace dam was not mentioned; indeed, the furnace appeared only in an omnibus item, "the old housing about the furnace with all the old iron and lumber."[50] By inference, the ironworks was a thing of the past. In all the legal papers only one ex-Hammersmith workman is identified, Thomas Tower, one of the Scots. The connection with the Undertakers' works is therefore mainly in the person of John Gifford, unless among the unknown workmen he assembled were other old employees of his at the Saugus and Monatiquot River plants. Whoever they may have been, they and their efforts stand submerged in a story where promotion, company organization, and lawsuits loom far larger than the making of iron.

At Rowley Village, now Boxford, near the Topsfield line, a bloomery was erected about 1670 that had more in common with the Raynham and Concord operations than with Gifford's. Like the former, it came into being by the initiative of local farmers and others who saw in a new industry both a source of profit and a stimulus to settlement. These men put up the money, found an "expert" to build the plant, and then leased it to him. The expert was Henry Leonard, in Gifford's day and perhaps in the Paine-Savage period a skilled forge hand at Hammersmith. Local historians have often claimed that this was "another" Leonard enterprise. Leonard had a sixteenth share in the business, which may have been given him as an inducement to take on the work. Ownership of the remainder was vested, however, in residents of Rowley Village, Topsfield, Salem Village, Ipswich, and other neighboring towns, some of them men of considerable substance. The limited available data suggest a quite understandable coming together of people with capital and a man with the appropriate technical skills.

Henry Leonard had become a freeman while still a resident of Lynn in 1668. By 1671–1672, when his name began to appear in court records as "manager of the Rowley Village ironworks," he was being referred to as "Mr. Leonard." It is possible that he owned a house

---

[49] This is another authorization to sell liquor to workmen, dated 30 June 1674. (*Essex R. & F.*, V, 372).

[50] Suffolk R. & F., 1912.

prior to 1673; by that date he was certainly leasing land. That he was running the business, contracting with colliers, and making and selling iron, all in his own name, is indisputable. The arrangements between an artisan who had obviously risen in the world, though still illiterate, and the ironworks shareholders are nowhere specified. It is not unlikely, however, that by the spring of 1673, and perhaps from the beginning, he was leasing the whole establishment at an annual rental of almost £190.[51]

Erected on land originally belonging to John Gould of Topsfield, the bloomery was a good one. A three-hearth forge with water-power-driven bellows and hammer, it made iron, including anchor stock, in fair quantity, and was a technological success. At a quite early date, however, signs of business weakness appeared. There were lawsuits over Leonard's failure to pay colliers according to agreement. Between the summer of 1672 and the spring of 1674 the ironworks manager was sued no less than a dozen times for debt, and at least twice for failure to deliver iron according to covenant.[52] Eventually, everything he owned was under attachment. With little question, Leonard had bitten off more than he could chew. In the fall of 1673, enmeshed in debts, he solved his problems by fleeing to New Jersey and another ironworks connection.

At the end of March, 1674, the owners of "Bromingum Forge," as the Rowley Village plant was called, met and decided to reclaim their property on the grounds that Leonard had taken off owing the better part of a year's rent and leaving the works in great danger of loss by fire. At the same time they agreed that they, or any two or three among them, were to "lett out" the forge *or* arrange to run the plant at their own charge, providing ore and charcoal and hiring workmen to "make Iron by the Tun." Re-entry was effected in April, with Mrs. Leonard delivering up her husband's lease in proceedings which, by the statement of the court officers involved, were wholly peaceful.[53]

The owners' preference would doubtless have been for an outright leasing of their forge. Seemingly, they were more than anxious to be

---

[51] *Essex R. & F.,* IV, 38; V, 131, 218–19, 271; VI, 239; Essex Deeds, VI, 36; Essex R. & F., XXI, 7–1. The total suggested in my text is based on the lower rental figure.

[52] *Essex R. & F.,* V, 77, 113, 130–32, 170, 173, 186, 196, 218–19, 227, 230, 285, 287–88.

[53] Essex R. & F., XXIII, 59–1; Suffolk R. & F., 1397.

wholly free of the Leonard family. Perhaps out of sheer necessity they fell back on arrangements which called for three of the partners, John Gould, Thomas Andrews, and Thomas Baker, or any two of them, to run the operation for the Company, and with Samuel, Nathaniel, and Thomas Leonard, Henry's sons, doing the actual work. True, the partners insisted that the Leonards bring the forge back to good repair, and reserved working and living space for at least one additional workman, possibly in order to have some on-the-spot supervision of Leonards, who were both young and wild. The sons, however, were to make iron, though on terms much different from their father's. They were to be paid at piece rates. The Company's representatives were to provide the wood, charcoal, and ore, receive and dispose of the iron, and allocate the profits. As things worked out, Gould and Baker took on the responsibilities and became, or came to be regarded as, the ironworks "clerks."[54]

By the summer of 1674 the Leonards had so managed to incur the wrath of the dominant group of shareholders that they announced that they would suspend operations altogether until more reliable workmen could be found. Mrs. Leonard greeted the news of impending dismissal with claims that the family was being most cruelly treated, and with prophecies of impending disaster for the plant, should it come to pass. God would right the wrong. There was a "trick" that workmen had to take care of those who turned them out of their jobs.[55] In the meanwhile, however, and with or without the approval of Gould and Baker, Nathaniel Leonard went ahead making iron according to his old covenant. Almost immediately one of Mrs. Leonard's dire prophecies was borne out. The forge was almost entirely destroyed by fire.

From the catastrophe, there developed litigation that carried on for nearly ten years. The main point at issue was whether or not John Gould had authorized Leonard to work despite the owner's instructions to the contrary. Meeting shortly after the fire, six of the owners, including Gould, authorized a new group, John Ruck, John Putnam, and Thomas Andrews, or any two of them, to take on workmen and get the works going once more. They also agreed, with Gould dissenting, that these men bring suit for damages. In September, Putnam

[54] Suffolk R. & F., 1397.
[55] *Essex R. & F.*, VI, 34, 54.

sued Gould, Baker, and Leonard, and won. An action of review in March, 1675, also went against Gould and left him responsible for damages of £210. He fared no better in still another action of review and in an appeal to the Court of Assistants which was heard in September. As late as 1684, however, Gould was still bringing actions of review in Essex County, and all to no avail.[56]

In all this, no evidence was offered that Nathaniel Leonard had pulled off the workmen's "trick," or set the fire, although there was not much doubt that he had stood by and watched it burn. At about the same period, however, there was still another fire that brought Thomas Leonard into court "upon suspicion of having a hand in burning the coal house at the Iron works." The magistrates found "great ground of suspicion," and Thomas was ordered not to come within seven miles of Topsfield or the ironworks on penalty of whipping. Thomas' subsequent career, which included a narrow escape from branding for highway robbery, can only suggest that he might well have been capable of the act of arson of which he had been suspected.[57]

The careers of other members of the Henry Leonard branch of the family were more than a little checkered; indeed, the activities of these Leonards illustrate more clearly than anything else the cleavage between ironworkers and Puritans. In time, some if not all of the brothers went to join their father in New Jersey, one of them clearly fleeing the avenging arm of justice as he did so.[58] Strangely enough, however, Henry, Samuel, Nathaniel, and Thomas were apparently all back in the Rowley Village area by 1679, and pledging good sums of money to the support of the minister at Topsfield at that. Perhaps, at long last, they, too, had been assimilated. The process was notably slower, and more strained, however, than it had been for their Taunton cousins.

The Leonards were not the only employees at Bromingum Forge with Hammersmith connections. When at one stage the owners insisted that Nathaniel not work the forge alone, and Samuel was not available, it was Thomas Look whom Nathaniel hired. When he

[56] *Ibid.*, V, 396; VI, 1–5, 34–35; Essex R. & F., XXXIX, 37–4; XL, 134–1, 2, 3; XLII, 79–1 to 87–6.

[57] *Essex R. & F.*, VI, 54, 256–59.

[58] *Ibid.*, VI, 56; Tinton Iron Works Papers (Monmouth County Historical Association), hereinafter referred to as Tinton Papers.

became wholly *persona non grata* with the owners, it was to John Flood, a forge hand who had learned his trade at the Lynn plant, that they turned. Eventually, John Vinton, the younger, became one of the chief workmen at the Rowley Village bloomery. His brother, Blaise, was also employed there, as were John Ramsdell, a collier, Thomas Wenmar, who had been at Hammersmith in Purchas' time, and Thomas Tower, the Scot whose connection with the Concord and Gifford's ironworks has already been mentioned. Geographical propinquity would have guaranteed that there be others, although some of the Bromingum work force seem to have had no connection at all with the older plant.[59]

The later years of the Rowley Village ironworks are as obscure as its beginnings. The court records are silent, which may mean that the plant was going smoothly, its employees at peace with themselves and the community, or was not going at all. The return of the Leonards prior to 1679 suggests that the works must then have been in operation. Local historians have set the termination of production at soon after 1680. When, in 1681, the town of Topsfield restricted the taking of bog ore to local residents, and set a fee of four pence a ton, payable to the town, Gould and an Ensign Peabody were authorized to mine twenty tons. Bromingum would have been its only conceivable destination. Equivalent privileges seem not to have been voted in later years, however, probably because the forge was no longer in business. Quite severe damage in the natural cutting of a new channel by the water which powered the plant, a catastrophe long remembered in Massachusetts, may well have been immediately responsible for its abandonment.

Shares in the ironworks changed hands no less than eleven times between March, 1671, and February, 1681. At no point is the full pattern of ownership clearly discernible, but by the latter date ten and a half sixteenths can be accounted for. Nearly half of the whole was then owned by John Ruck, a Salem merchant, mainly by purchase from Gould. Gould, Jonathan Wade of Ipswich, and Simon Bradstreet of Andover held at least three shares. Title to the remainder, which probably represented original investment in the

---

[59] *Essex R. & F.*, VI, 3, 35; Suffolk R. & F., 1397. Considerable "visiting" back and forth between the Rowley Village works and Hammersmith makes positive identification of workers with one plant or the other more than a little difficult.

partnership, seems not to have changed hands in transactions of record. So far as surviving deeds and wills go, the fate of all of the shares is as vague as that of the physical plant that they represented. Since, with the exception of Nathaniel Putnam, who seems to have been allowed to keep the £210 he recovered from Gould, no one of the partners seems to have profited from the ironworks, it is probable that the owners merely wrote off their shares as bad investment.

At some point soon after 1680, then, a quite ambitious locally sponsored ironworks ground to a halt. The reasons for failure are impossible of assessment. The scale of operations seems to have been sound. There is no evidence of a running out of ore. The water damage, great as it may have been, ought to have been susceptible of repair if the plant were making iron and disposing of it at a profit. Labor costs must have been high, and workmen hard to keep, but with the windup of Hammersmith there ought to have been some easing of the labor problem. There was obviously conflict among partners who had to depend on others' technical capacities. It is probable that capitalization had been inadequate from the beginning, possible that management was poor. This, like the other plants, had to face the competition of imported iron. Whatever the cause or causes, Bromingum Forge was, with little question, just another early American ironworks failure.

The most remote of the plants to which Hammersmith people moved on was, in its heyday, a most impressive enterprise. This was the ironworks at Tinton Falls near the present town of Red Bank, Monmouth County, New Jersey, but in its day within the limits of Shrewsbury, a town settled by Connecticut people about 1664. Begun on a modest scale, about 1674, by James Grover and some unidentified partners, it eventually grew into a heavily capitalized iron plantation, part of the more than 6000-acre Manor of Colonel Lewis Morris, and replete with land subsidies, tax exemptions, and freedom of workers from military service.[60] Resident there, and employed during both pioneering and large-scale phases, were Henry Leonard and his sons, fresh from their difficulties at Rowley Village.

How word of ironmaking projects in New Jersey reached Henry Leonard is not clear. He may have been solicited for his trade skills.

[60] Dean Freiday, "Tinton Manor: the Iron Works," *N. J. Hist. Soc. Proc.*, LXX (1952), 250–61.

He may have put out feelers for a new connection offering better economic opportunity and, perhaps, fewer social constraints. The Leonards may have had acquaintances among the Shrewsbury settlers. In any event it is probable that Leonards built the forge at Tinton Falls, conceivable that they were Grover's "partners." Given the mountainous debts in their immediate background, one must assume that if they were part owners, their shares had been given them as an inducement to their coming. They had a house, probably built for them by Grover and his associates. At an early date, too, they came to own no less than 1170 acres of land, purchased from the Indians, laid out by warrants issued by Governor Carteret, and subsequently confirmed by the Proprietors of the Eastern Division of New Jersey. Obviously, even without a share in the proprietorship, things were looking up for the Leonards.

The ownership picture is obscure even in the case of men of wealth known to have been interested in the venture. In December, 1674, Grover mortgaged "an equall great part" of the business to Cornelis Steenwyck, a rich New York merchant.[61] A year later Grover sold outright to Colonel Morris enough of the parts or shares which he held in fee simple to make up a full half interest. Whether or not this was one and the same "half" is not indicated. In two instances, data on Morris' purchase of other men's shares are available. We also know, however, that at least in the early stages of his operations, Morris had partners.[62] Eventually, the works came to be known as Morris', but when and how he assembled all of the fractional interests is not clear. His ownership of even a half of the ironworks was, of course, enough to guarantee adequate financial support of the undertaking.

The Colonel's coming into the business was also good for the Leonards. Whatever their original status may have been, they appear in accounts covering the Morris phase of operations as highly remunerated employees. They seem also to have done well in the sale and lease of lands they had acquired from the Indians. Early in 1676, Henry Leonard sold Morris five of many acres and granted him ore and timber rights in the remainder, with fees to be set by "3 honist Naibors."[63] Samuel Leonard sold outright a large tract of land in

[61] Freiday, *op. cit.*, 252, and n. 7.
[62] Tinton Papers.

1676. By any test, the family ought to have been faring not only better than they had at Rowley Village, but better than the workers at seventeenth-century colonial ironworks in general.

When Morris bought Grover's half interest, the enterprise probably consisted of a still incomplete forge and outbuildings, watercourses, and land. Eventually he had an ironworks consisting of a forge that included two fineries, a chafery and a power-driven hammer, and, one would assume, a blast furnace.[64] There were also a corn mill, a dwelling house and stable, the Leonards' old house, a smith's forge, and accommodations, separate, for white workmen and Negro slaves, at least sixty of whom worked for Morris, some of them apparently right at the ironworks. The cost of free labor was high. There were twenty-five men who had been employed for seven years at an annual wage of £20.[65] Managerial and clerical salaries and the wages of highly skilled people like the Leonards and Grover ran higher still. By the first part of 1684 no less than £8680 had been poured into the ironworks.[66]

Production data are nonexistent; so is the credit side of the ironworks accounts. It is therefore impossible to estimate the degree of success of the operation at Tinton Falls. One would infer from the record of expenditures, however, that activity had been vigorous and on a large scale. Certainly there was a quite considerable geographical range in the Company's transactions. Not only were there trips to and from New York but to New Haven and Taunton as well, now to freight in the goods sold to workers, now to secure cast iron and, perhaps, new recruits for the labor force.

Tinton Ironworks is also interesting in terms of legal status and relationship to the government of New Jersey. In the first place, the plant was part of a "Manor" which had its own petty court and carried its share as a separate jurisdiction in joint township-financed

---

[63] According to the deed, the land was adjacent to the ironworks. Leonard excepted from the lease of timber rights such wood and timber as his "vocation" required. This might imply that he was by then carrying on more or less independent operations. The all but complete absence of his name from the accounts would appear to support such an inference.

[64] This version of the ironworks' working units is inferred from various entries in the accounts, especially in a folio summarizing Grover's credits over the years.

[65] Freiday, *op. cit.*, 258, and n. 45.

[66] Tinton Papers.

activities, such as bridge building.[67] None of the New England iron-works was independent of regular town governments. Furthermore, the grants and privileges which it enjoyed were sweeping. In November, 1675, prior to his purchase from Grover, Morris petitioned for, and eventually received, a seven-year exemption of land and workers from taxation, barring a state of war; five years' freedom from rent; the exemption of workers from arrest for debt, although not from civil suits, and from regular militia duty. Besides huge land grants—a tract of 3450 acres, one of 150 and another of 500—Morris was awarded ore-digging privileges over a wide area in which some private property was included. In the latter, the owners were to be paid damages. Otherwise, however, the Colonel and his associates were free to take what they wanted wherever they found it. Clearly this ironworks was no struggling orphan.

As at Hammersmith, operations can be precisely dated only for the years covered in surviving accounts. The latest date appearing in the Tinton Falls business papers is February 2, 1684, the day on which the debtor account stood at £8680. Evidence of activity there-after is thin and subject to conflicting interpretations. Oldmixon, writing in 1708, suggested that the ironworks was still standing, but added that he understood that it had not been "any great Benefit to the Proprietors."[68] The Morris Papers at Rutgers carry detailed references on all kinds of activities at Tinton Manor in later years but data on the making of iron are conspicuously lacking. Although Lewis Morris, the Colonel's nephew and heir, expressed interest in measures aimed to benefit the manufacture of iron in New Jersey in 1741, his will, proved in 1746, made no mention of the ironworks proper in its conveyance of the Manor and its "privileges," liberties, mines and minerals."[69] By inference, the plant was by then abandoned but neither the why nor the when are presently ascertainable.

Though it is only in the Leonards that Tinton Ironworks and Hammersmith seem to have been related, to the historian they have much in common. Both were large. Both must have been set up with an eye toward future population growth and increased demand for

[67] Freiday, *op. cit.*, 255, and n. 24.

[68] John Oldmixon, *The British Empire in America* (2 vols., London, 1708), I, 138.

[69] Charles S. Boyer, *Early Forges and Furnaces in New Jersey* (Philadelphia, 1931), 199.

their products. Both seem clearly to have been business failures. Within the limits of presently available data the first ironworks in New Jersey, like its Massachusetts counterparts at Lynn and Braintree, stands as symbol of the vigor of early colonial metallurgical ambitions, of the spirit of enterprise in industry. That Tinton Falls achieved its grand scale wholly without recourse to English capital gives it a special place in our history.

One ironmaking operation with indirect ties to Hammersmith appears to have left but few traces in published records. This was the forge at Pawtucket, Rhode Island, erected by Joseph Jenks, Jr., about 1672. His efforts with a sawmill and slitting mill at Concord clearly unsuccessful, this son of an able father removed to the younger colony in the latter part of 1668. It was to work with Rhode Island's timber resources that he first intended to turn, or so one would infer from a grant of timber-cutting privileges on Pawtuxet River from the proprietors of Warwick in March of 1669.[70] The practical outcome of this grant is obscure. In the fall of 1671, however, Jenks bought sixty acres of ground near Pawtucket Falls and there erected, sooner or later, both a sawmill and a forge. To those were added, in time, a gristmill, a coal house, smith's shops, and perhaps another sawmill as well. Of the forge, the unit in which we are directly interested, it is possible only to say that it was destroyed in King Philip's War but was rebuilt, probably not before 1684 but certainly by 1688; that it served as a training ground for Joseph's brother, Daniel, and at least one son, Nathaniel; and that it was still in business in 1721 when Nathaniel bequeathed his share to a son of the same name. That the total operation had been at least moderately successful seems to follow from the Rhode Island Jenks' rise in the world.[71]

Here and there in the old records one comes upon references to ironmaking and iron-fabricating activities in which one might expect that sons or grandsons of the pioneer New England ironworkers had been employed. Between 1682 and 1710, for example, John Hubbard, a prosperous Boston merchant, appears to have set up at least an embryonic iron "empire" in Massachusetts. On Monatiquot River

---

[70] John Osborne Austin, *The Genealogical Dictionary of Rhode Island; Comprising Three Generations of Settlers Who Came before 1690* (Albany, 1887), 112.

[71] *Early Records of Providence,* IV, 6–7; XV, 189, 209, 212; XVI, 251; XVII, 14, 119, Savage, *Genealogical Dictionary,* II, 543.

in Braintree, and in the general vicinity of the old forge of the Undertakers, he had a blast furnace and, for a while, no less than two forges.[72] At Newton, at the Lower Falls of the Charles, he had a two-hearth forge and a power hammer. We know a little of his purchase, mortgage, and sale of portions of his property. We may infer from the fact of his dying insolvent that his interest in iron had not paid, and this despite vigorous entrepreneurial efforts. Except for the circumstance of a Vinton's winding up as owner of the Braintree millsite, however, there seems to be nothing whatever on which to base a claim of continuity from the first Massachusetts ironworks. The identity of Hubbard's own workmen is wholly obscure. We do not know whether any of those who toiled for him were heirs of the Hammersmith training legacy. The last strands of that thread, here and doubtless elsewhere, are lost to history.[73]

What may we conclude from those parts of the story for which we have data, however fragmentary, however tantalizing? There is, first, good reason to regard most of these seventeenth-century plants as business failures. With but few exceptions, the immediate heirs of Hammersmith and Braintree Works fared little better than those first enterprises. To avoid the final judgment that the times were not ripe is impossible. There was ore. There was a market. There were the technical skills and the technical facilities. There was active interest and active effort from entrepreneurs, from "government," and from communities at large. There were variations in scale of enterprise, in techniques, and in business forms and procedures put through by people whose situation ought to have guaranteed realistic estimates of business potentials and sound managerial controls. On the other side of the ledger, however, were inadequate capitalization, high wage bills, and, above all, the competition of imported iron. It was to be well into the eighteenth century that the colonial American iron industry was to come into its own, and then, for the larger part, outside of New England. If most of the ironworks here chronicled were failures, some of them were magnificent failures. If most of them went "back to Nature," the spirit of enterprise which had pro-

---

[72] Hubbard's total efforts, strangely neglected by local historians, deserve far more extended treatment than is possible here.

[73] While one would expect some of the Hammersmith people to have gone to Hubbard's works, I have found no data to substantiate the inference.

duced them did not die out. Though few prospered financially from them, they rewarded the total community less, perhaps, with iron than as socioeconomic catalysts. What they inherited from Hammersmith and Braintree Works and other early business operations, the seen and the unseen, the direct and the oblique, the verifiable and the imponderable, became part of the stuff of which men were to build an American industrial civilization.

Bibliographical Note

THE AVAILABILITY in a number of places of convenient listings of books on the iron industry and of materials on the American colonies has led me to dispense with a formal bibliography in favor of full citation in the notes for my first mention of a given pamphlet, book, compilation, or serial publication. Thereafter I cite by author and brief title, except within the limits of single chapters, where I use the standard abbreviations. Exceptions are certain items of which I have made widespread use and which are indicated throughout by brief title or abbreviation. Those in published form include: Nathaniel B. Shurtleff, ed., *Records of the Governor and Company of Massachusetts Bay* (Boston, 5 vols. in 6, 1853–54), cited as *Mass. Records; Suffolk Deeds* (Boston, 14 vols., 1880–1906), cited by title; George Francis Dow, ed., *Records and Files of the Quarterly Courts of Essex County, Massachusetts* (Salem, 8 vols., 1911–21), cited as *Essex R. & F.;* and *Winthrop Papers* (Boston, 5 vols., 1929–47), also cited by title.

Since the Essex County materials were published in summary, and since the chronological limits of my study exceed those of Mr. Dow's volumes as well as those of the *Winthrop Papers* presently in print, I have had recourse in both instances to the original MSS. Impressed with the calibre of the W.P.A. verbatim transcript of the former, available in the office of the Clerk of Courts, I have used its volume and page numbering in my annotation. Whenever "Essex R. & F." appears without italics, reference is to this typed transcript. (On a few occasions I have added the word "typescript" to prevent confusion as I coupled citations of published and typescript versions in a single note.) Similarly, with the voluminous Winthrop Papers in the Massachusetts Historical Society and the Suffolk Deeds I have differentiated between published and MSS. data by italicizing in the case of the former.

Another major source, "A Collection of Papers Relating to the Iron Works at Lynn and More Particularly to the Suit between Mr. John Gifford, the Agent for the Undertakers of the Iron Works, and the Inhabitants of the Massachusetts Bay Colony, Dated 1650 et seq.," in the Baker Library of the Harvard Business School, is also available both in typescript and in MSS. In my notes I have cited the collection as "Ironworks Papers" and have used the pagination of the typescript, occasionally pointing out corrected readings, as in the case of the Essex Court and Winthrop materials.

References to the Essex Deeds, those of Suffolk of later date than are available in published form, and the Suffolk Probate Records are in terms of the clear clerk's copies on file in the appropriate offices. Materials in the State Archives are cited as "Mass. Arch." The records of the quarterly courts of Middlesex County, used in the originals, although on a limited basis, are indicated as "Middlesex R. & F." Finally, and in minor deviation from common practice, I have chosen to cite as "Suffolk R. & F." materials in the archives of the Supreme Judicial Court of Massachusetts, and to indicate only the case or docket number.

The English MSS. sources, which were provided by Mr. B. W. Clapp, are given in full except for papers in the Public Record Office. These are indicated by abbreviation as follows: Chancery—C; Exchequer—E; High Court of Admiralty—H. C. A.; and State Papers—S. P.

Index

309